Zirconia and Innovative Biomaterials for Dental and Biomedical Applications

Zirconia and Innovative Biomaterials for Dental and Biomedical Applications

Editors

Sangwon Park
John G. Fisher

MDPI • Basel • Beijing • Wuhan • Barcelona • Belgrade • Manchester • Tokyo • Cluj • Tianjin

Editors
Sangwon Park
Department of Prosthodontics
Chonnam National University
Gwangju
Korea, South

John G. Fisher
School of Materials Science
and Engineering
Chonnam National University
Gwangju
Korea, South

Editorial Office
MDPI
St. Alban-Anlage 66
4052 Basel, Switzerland

This is a reprint of articles from the Special Issue published online in the open access journal *Materials* (ISSN 1996-1944) (available at: www.mdpi.com/journal/materials/special_issues/Zirconia_Biomaterials_Dental).

For citation purposes, cite each article independently as indicated on the article page online and as indicated below:

LastName, A.A.; LastName, B.B.; LastName, C.C. Article Title. *Journal Name* **Year**, *Volume Number*, Page Range.

ISBN 978-3-0365-7815-6 (Hbk)
ISBN 978-3-0365-7814-9 (PDF)

Contents

About the Editors . . . . **vii**

Preface to "Zirconia and Innovative Biomaterials for Dental and Biomedical Applications" . . **ix**

Kumaresan Sakthiabirami, Vaiyapuri Soundharrajan, Jin-Ho Kang, Yunzhi Peter Yang and Sang-Won Park
Three-Dimensional Zirconia-Based Scaffolds for Load-Bearing Bone-Regeneration Applications: Prospects and Challenges
Reprinted from: *Materials* **2021**, *14*, 3207, doi:10.3390/ma14123207 . . . . **1**

Mee-Jin Jun, Jin-Ho Kang, Kumaresan Sakthiabirami, Seyed Aliakbar Hosseini Toopghara, Ye-Seul Kim and Kwi-Dug Yun et al.
The Impact of Particle Size and Surface Treatment of Zirconia Suspension for Photocuring Additive Manufacturing
Reprinted from: *Materials* **2023**, *16*, 1670, doi:10.3390/ma16041670 . . . . **35**

Jin-Ho Kang, Kumaresan Sakthiabirami, Hyun-Ah Kim, Seyed Aliakbar Hosseini Toopghara, Mee-Jin Jun and Hyun-Pil Lim et al.
Effects of UV Absorber on Zirconia Fabricated with Digital Light Processing Additive Manufacturing
Reprinted from: *Materials* **2022**, *15*, 8726, doi:10.3390/ma15248726 . . . . **49**

Mariana Novais, António Sérgio Silva, Joana Mendes, Pedro Barreiros, Carlos Aroso and José Manuel Mendes
Fracture Resistance of CAD/CAM Implant-Supported 3Y-TZP-Zirconia Cantilevers: An In Vitro Study
Reprinted from: *Materials* **2022**, *15*, 6638, doi:10.3390/ma15196638 . . . . **59**

Seon-Ki Lee, Min-Kyung Ji, Yu-Jin Jo, Chan Park, Hoonsung Cho and Hyun-Pil Lim
Effect of Non-Thermal Plasma Treatment of Contaminated Zirconia Surface on *Porphyromonas gingivalis* Adhesion and Osteoblast Viability
Reprinted from: *Materials* **2022**, *15*, 5348, doi:10.3390/ma15155348 . . . . **71**

Oskar Bunz, Marie-Christine Steegmann, Korbinian Benz, Holger Testrich, Antje Quade and Ella A. Naumova et al.
Human Gingival Fibroblast Adhesion and Proliferation on Hydroxyapatite-Coated Zirconia Abutment Surfaces
Reprinted from: *Materials* **2022**, *15*, 3625, doi:10.3390/ma15103625 . . . . **89**

Flavia Roxana Toma, Mihaela Ionela Bîrdeanu, Ion-Dragoș Uțu, Roxana Diana Vasiliu, Lavinia Cristina Moleriu and Liliana Porojan
Surface Characteristics of High Translucent Multilayered Dental Zirconia Related to Aging
Reprinted from: *Materials* **2022**, *15*, 3606, doi:10.3390/ma15103606 . . . . **101**

Jae-Gon Jang, Jin-Ho Kang, Kwang-Bum Joe, Kumaresan Sakthiabirami, Kyoung-Jun Jang and Mee-Jin Jun et al.
Evaluation of Physical Properties of Zirconia Suspension with Added Silane Coupling Agent for Additive Manufacturing Processes
Reprinted from: *Materials* **2022**, *15*, 1337, doi:10.3390/ma15041337 . . . . **123**

Waqas Ahmed Sarwar, Jin-Ho Kang and Hyung-In Yoon
Optimized Zirconia 3D Printing Using Digital Light Processing with Continuous Film Supply and Recyclable Slurry System
Reprinted from: *Materials* **2021**, *14*, 3446, doi:10.3390/ma14133446 . . . . **137**

Felice Lorusso, Sammy Noumbissi, Inchingolo Francesco, Biagio Rapone, Ahmad G. A. Khater and Antonio Scarano
Scientific Trends in Clinical Research on Zirconia Dental Implants: A Bibliometric Review
Reprinted from: *Materials* **2020**, *13*, 5534, doi:10.3390/ma13235534 . . . . **149**

About the Editors

Sangwon Park

Sang-won Park is a Professor and Head of the Department of Prosthodontics at the School of Dentistry, Chonnam National University, and also the director of the Biomedical Evaluation and Research Center, at the School of Dentistry, Chonnam National University. He obtained his M.S. (1988) and Ph.D. (1995) from the School of Dentistry, Chonnam National University. He was the Chairman of the Department of Prosthodontics and general director of Chonnam National University Hospital (2003–2011). His research deals with the design and development of zirconia and bio-ceramic 3D printing, implant surface modifications, and dental and biomaterials for clinical applications.

John G. Fisher

John Fisher is an Associate Professor in the School of Materials Science and Engineering at Chonnam National University. He graduated from the Department of Engineering Materials, University of Sheffield, United Kingdom, with a Ph.D. in July 1999. He worked as a postdoctoral researcher at the Korea Institute of Energy Research from 1999 to 2001, studying the reaction bonding of Si3N4. He then moved to the Korea Advanced Institute of Science and Technology (KAIST) from 2001 to 2005 as a postdoctoral researcher and visiting professor. At KAIST, he worked on single crystal growth of Pb(Mg1/3Nb2/3)O3–35 mol% PbTiO3 and abnormal grain growth behaviour in BaTiO3. From 2005 to 2007, he was a postdoctoral researcher at the Jožef Stefan Institute, Slovenia, studying single crystal growth of (K0.5Na0.5)NbO3. He returned to KAIST in 2007 and continued studying (K0.5Na0.5)NbO3. In July 2010, he became an Assistant Professor in the School of Materials Science and Engineering at Chonnam National University. His research interests are the processing and characterization of ceramics and single crystals for lead-free piezoelectric applications. He has authored or co-authored 54 SCI(E) manuscripts and 2 book chapters.

Preface to "Zirconia and Innovative Biomaterials for Dental and Biomedical Applications"

The use of zirconia in the biomedical field was initiated in the early 1970s, and the application of zirconia as a prosthetic material in dental-related applications was unlocked in the 1990s. Very recently, many scholars established the use of zirconia in the dental and biomedical field as an implant and as a scaffold. Thus, zirconia and zirconia-based materials have a wide range of applications in the biomedical and dentistry fields owing to their excellent mechanical properties, aesthetics, and biocompatibility. This Special Issue critically explores the art and state of zirconia surface treatments (mechanical/chemical/physical), which are a significant challenge in implantology in designing implant biomaterial using advanced technologies which have evolved rapidly to enrich the biological and osteointegration process of dental implants. The surface characteristics are proposed to improve the capacity of anchorage into the bone which determines the long-term clinical success rate. In addition, digital technology (CAD/CAM and 3D printing) in dentistry plays a crucial role in the fabrication of dental restorations and prostheses because of its efficient manufacturing process with high accuracy in a short time. The perfectibility of technology and the suitability of materials are considered the ultimate requisite for the future.

With an increase in age-related pathologies and associated illnesses worldwide, there is a necessity for the advancement of biomaterials to substitute tissue loss and boost regenerative mechanisms, as well as to ensure the healing process. Bioceramics, polymers, and metals are such materials that have been used to repair and restore bone and dental defects, especially for the reconstruction/regeneration of hard tissues. To achieve three-dimensional tissue regeneration, the porous scaffold with multifunctional properties was fabricated using additive manufacturing and conventional techniques.

Sangwon Park and John G. Fisher
Editors

materials MDPI

Review

Three-Dimensional Zirconia-Based Scaffolds for Load-Bearing Bone-Regeneration Applications: Prospects and Challenges

Kumaresan Sakthiabirami [1,†], Vaiyapuri Soundharrajan [2,†], Jin-Ho Kang [1], Yunzhi Peter Yang [3] and Sang-Won Park [1,*]

1 Department of Prosthodontics, Dental Science Research Institute, School of Dentistry, Chonnam National University, Gwangju 61186, Korea; sakthikarthi.dentist@gmail.com (K.S.); jhk.bme1002@gmail.com (J.-H.K.)
2 Department of Materials Science and Engineering, Chonnam National University, Gwangju 61186, Korea; soundharajan.007@gmail.com
3 Department of Orthopaedics Surgery, Stanford University, Stanford, CA 94305-5110, USA; ypyang@stanford.edu
* Correspondence: psw320@jnu.ac.kr; Tel.: +82-62-530-5842; Fax: +82-62-530-5634
† These two authors equally contributed.

Abstract: The design of zirconia-based scaffolds using conventional techniques for bone-regeneration applications has been studied extensively. Similar to dental applications, the use of three-dimensional (3D) zirconia-based ceramics for bone tissue engineering (BTE) has recently attracted considerable attention because of their high mechanical strength and biocompatibility. However, techniques to fabricate zirconia-based scaffolds for bone regeneration are in a stage of infancy. Hence, the biological activities of zirconia-based ceramics for bone-regeneration applications have not been fully investigated, in contrast to the well-established calcium phosphate-based ceramics for bone-regeneration applications. This paper outlines recent research developments and challenges concerning numerous three-dimensional (3D) zirconia-based scaffolds and reviews the associated fundamental fabrication techniques, key 3D fabrication developments and practical encounters to identify the optimal 3D fabrication technique for obtaining 3D zirconia-based scaffolds suitable for real-world applications. This review mainly summarized the articles that focused on in vitro and in vivo studies along with the fundamental mechanical characterizations on the 3D zirconia-based scaffolds.

Keywords: 3D zirconia-based scaffold; bone-regeneration applications; composite; coating; fabrication techniques; bioceramics

Citation: Sakthiabirami, K.; Soundharrajan, V.; Kang, J.-H.; Yang, Y.P.; Park, S.-W. Three-Dimensional Zirconia-Based Scaffolds for Load-Bearing Bone-Regeneration Applications: Prospects and Challenges. *Materials* **2021**, *14*, 3207. https://doi.org/10.3390/ma14123207

Academic Editor: Montserrat Colilla

Received: 11 May 2021
Accepted: 1 June 2021
Published: 10 June 2021

1. Introduction

1.1. General Overview of Zirconia Bioceramics

Zirconium dioxide (zirconia) was first discovered by the German chemist Martin Heinrich Klaproth in 1789 [1]. The use of zirconia in the biomedical field emerged in 1969, as it is a promising alternative to alumina and metal for the construction of hip prosthesis in orthopaedic applications [2]. Zirconia is an oxide form of zirconium (strong transition metal), which does not exist in pure form in the Earth's crust. It is found in the minerals baddeleyite and zircon. Zirconia is a polycrystalline ceramic that exhibits three different crystallographic phases depending on the temperature and pressure: monoclinic (M), tetragonal (T) and cubic (C) [3]. Pure zirconia with a monoclinic structure is stable at room temperature and up to 1170 °C and has inferior mechanical properties to the other two phases. The transition from the monoclinic phase to the tetragonal phase occurs between 1170 °C and 2370 °C, accompanied by approximately 4–5 vol% reduction. Zirconia shrinks to the cubic phase above 2370 °C and up to the melting point (2680 °C) [4]. The transformation of the tetragonal zirconia lattice into the monoclinic phase occurs with approximately 3–4% volume expansion upon cooling. This phase-transformation behaviour of zirconia results in crack propagation over time, because of the internal stress produced during cooling. The aforementioned phenomena can be inhibited by the addition of a relative amount of metallic oxides (also known as 'dopants' or 'stabilising oxides') such as Y_2O_3, MgO, CaO and CeO_2 [5]. This doped zirconia exhibits a unique property known as 'transformation toughening', and it is considered a key advantage for biomedical applications in orthopaedics and dentistry [5,6].

1.2. *Inevitability of Widespread Use of Zirconia Bioceramics in Biomedical Applications*

It is well known that zirconia is available in various chemical forms; however, in the field of biomedical research, only three types have primarily been used: yttrium-stabilised tetragonal zirconia polycrystals (3Y-TZP), magnesium-doped partially stabilised zirconia (Mg-PSZ) and zirconia-toughened alumina (ZTA) are the major contributors in biomedical and dental applications [7]. The use of zirconia bioceramics in dental applications in the form of dental prostheses started in the early 1980s and has attracted considerable attention in the dental community [5,8]. In the early days, wide variety metal alloys were employed in dental applications; titanium alloys exhibited clinical success rates of 92–98% yet had minor shortcomings corrosion induced metal ion dissolution into body fluids. [9,10]. The arrival of zirconia bioceramics was a blessing to dentistry owing to their tooth-like colour, high fracture toughness and low-temperature conductance. Owing to its high flexural strength (900 MPa) and non-reactivity with body fluids, researchers successfully employed zirconia for crown and bridge applications [11,12]. Extensive research attempts have been made to utilise the mechanical and biological advantages of zirconia in the form of dental posts as a synthetic tooth root to replace missing teeth [13,14]. Numerous reviews of the use of zirconia for implant applications have been conducted and served as a potential reference for the dental and orthopedic research community [3,7,9,10].

More importantly, after the effectual utilisation of zirconia ceramics for the construction of tooth reinforced repairs, clinicians extended the use of the valuable features of zirconia ceramics (that is, the lower elastic modulus and higher toughness for implant-reinforced renovations) [15]. After the commercialisation of zirconia dental implants in 1987 by Sigma Implants (Sandhause, Incermed, Lausanne, Switzerland), zirconia implants became widely accessible. However, the use of zirconia scaffolds for comprehensive load-bearing applications is commercially nonviable owing to various challenges, such as scaffold fabrication and surface modification. The promising development of zirconia implants has been documented worldwide, and several reviews have been conducted in recent years. For instance, Soon et al., and Yin et al., summarised the recent advances in fabrication techniques for zirconia implants [16,17]. Cionca et al., and Amleh et al., compared the clinical advantages and difficulties between zirconia and titanium implants in the previous review [18,19]. These reviews provide a basic understanding of the origin and evolution of zirconia bioceramics from dental prostheses to biomedical implant applications.

1.3. *Commencement of Zirconia over Calcium-Phosphate Scaffolds in Bone-Regeneration Applications*

Bone-tissue regeneration using synthetic biomaterials with an identical chemical composition to human bone has developed as a pioneering and favourable strategy for the restoration of bone defects [20]. Calcium phosphate (CP) has chemical and biological behaviours identical to those of natural bone, and calcium phosphate-based materials have been extensively studied by various research groups worldwide [21–23]. However, they have their own faintness in terms of mechanical properties, which is a major concern for load-bearing applications [24]. More importantly, CP-based materials are biodegradable in nature when subjected to the human body and fail to support the reconstruction process because they do not maintain their original shapes [25]. To overcome this calamity, researchers used a composite formation strategy, blending calcium phosphate with mechanically strong and biologically inert zirconia [26]. It is well known that zirconia has an elastic modulus, fracture toughness and osseointegration properties similar to those of human bones [27]. Since traditional two-dimensional (2D) biomaterials cannot retain three-dimensional (3D) architectures, 2D designs have a limited ability to mimic the multi-dimensional extracellular background, which is essential for promoting cell feasibility and functionality [28]. The design of synthetic bone-graft materials in the form of 3D porous scaffolds loaded with tissue-activating features or precise cells to launch bone restoration is an innovative approach [29,30].

In recent times, considerable research attention has been focused on the progress of engineering techniques for developing CP biomaterials in the form of 3D scaffolds [31,32]. In a recent review, Ryan et al., reported that the available 3D-printing techniques have matured sufficiently to formulate 3D porous CP scaffolds [33]. Additionally, they stated that 3D-printed CP scaffolds must reach the level 1 preclinical stage to confirm effectiveness in a large animal before bone-regeneration checks are performed for human trials. However, the synthesis of 3D zirconia-based scaffolds remains in a stage of infancy because of the lack of availability of 3D fabrication techniques [34].

Few 3D fabrication techniques have been successfully investigated for fabricating 3D zirconia-based scaffolds.

The evolution of the fabrication techniques for the fabrication of 3D zirconia-based scaffolds over time is presented in Scheme 1. As shown, 3D zirconia-based scaffold fabrication via the sponge replica technique for bone-regeneration application started in the late 2000s. However, the use of modern 3D fabrication techniques for 3D zirconia-based scaffolds started very recently. According to the types of 3D fabrication techniques, this article can be divided into three main sections: conventional, hybrid and digital (Scheme 2).

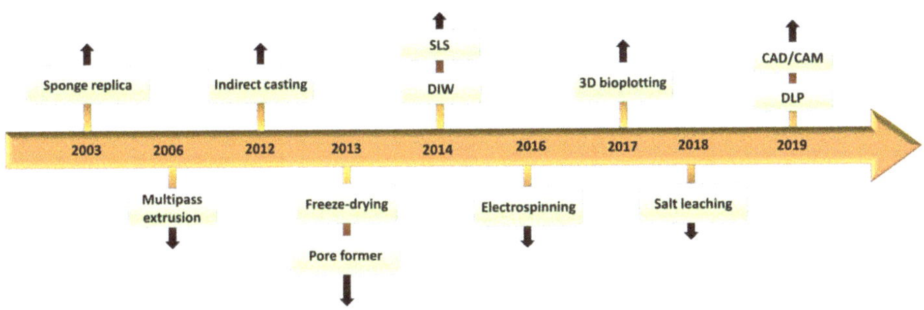

Scheme 1. Evolution of 3D zirconia-based scaffolds (with regard to the scaffold fabrication technique) for bone-regeneration applications.

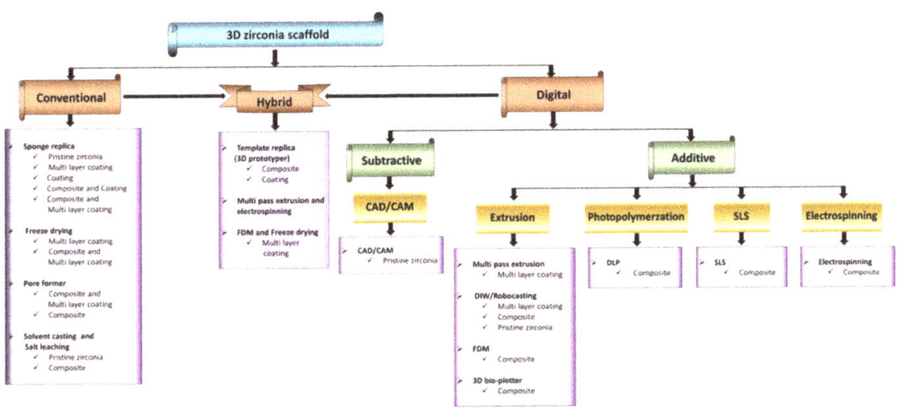

Scheme 2. Overview of the 3D fabrication techniques for zirconia-based scaffolds, along with the surface optimisation tactics using coating and composite strategies.

Originally, researchers investigated zirconia-based scaffolds using conventional moulding approaches, including sponge replicas, freeze-drying, pore-formers and salt leaching. Even though the conventional techniques for fabricating zirconia-based scaffolds are economical, the precise formation of complex structures is difficult [34]. Thus, in recent years, researchers have used 3D digital techniques as alternative methods for fabricating zirconia-based scaffolds, as for other biomaterials. In the following sections, we discuss the roles of previously reported composite/coating approaches in enhancing the mechanical and biological properties of zirconia-based scaffolds. The most common composite and coating materials used for the bioengineering of zirconia-based scaffolds are presented in Scheme 3.

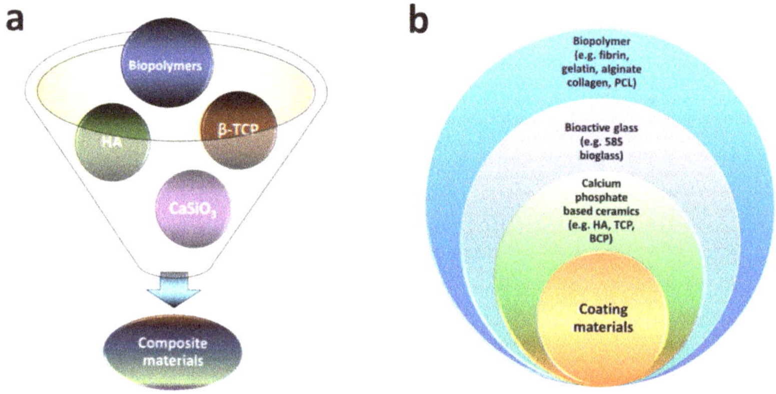

Scheme 3. (**a**) Composite materials and (**b**) coating materials for bioengineered zirconia-based scaffolds.

Thus, this review systematically summarises the challenges and advancements in the development of 3D fabrication techniques for 3D zirconia scaffolds and their role in bone-regeneration applications for the first time as per the author's knowledge. We focused on studies in which the biological activity of zirconia scaffolds was successfully demonstrated in vitro and in vivo. Finally, guidance for future research directions for formulating optimal zirconia scaffolds via conventional and modern 3D digital techniques is provided. More importantly, we comprehensively addressed the gap between the academic reliability and clinical reliability of various 3D zirconia scaffold techniques for bone regeneration. We hope that this review will draw attention to the production of 3D zirconia scaffolds and promote their clinical implementation.

2. 3D Zirconia-Based Scaffolds via Conventional Technique

2.1. Sponge Replica Technique

The polymeric sponge replica method is the most popular technique for producing interconnected porous bioceramics. This method is known for its simplicity. It involves the coating of open-cell polymeric foam with the desired bioceramic slurry and subsequent thermal treatment to burn-out the polymeric foam and obtains a bioceramic with a 3D structure similar to that of the original polymeric foam [35]. The composite/coating design mixture of biocompatible materials and bioinert porous zirconia-based scaffolds have been studied extensively by multiple research teams [36]. By fixing bioinert zirconia scaffolds as a mainstream loadbearing framework, several scholars have fabricated advanced coating/composite formations using biocompatible and osteoinductive bioceramics.

Hydroxyapatite ($Ca_{10}(PO_4)_6(OH_2)$, HA) is a well-known material in the calcium-phosphate family owing to its analogous properties to human bone components and has been employed for bone replacement [37]. Tricalcium phosphate (β-TCP), which is commonly referred to as bone ash, has a chemical formula of $Ca_3(PO_4)_2$. Similar to HA, TCP is rich in calcium and phosphorus, which can induce new-bone construction [38]. β-TCP has been formulated directly in the form of scaffolds and tested for bone-tissue regeneration applications [39]. Similarly, the biphasic calcium phosphate (BCP) bioceramic—a mixture of TCP ($Ca_3(PO_4)_2$) and HA ($Ca_{10}(PO_4)_6(OH_2)$—has been used as a bone-craft material because it has a good chemical bone-bonding property and a greater bioresorption capability than HA and TCP individually [40]. BCP promotes the growth of osteoblasts/osteoclasts in a more habitual fashion than pristine HA and TCP [41]. Our literature assessment indicated that HA, TCP and BCP are the calcium-phosphate ceramics that are predominantly employed to formulate bioengineered zirconia scaffolds. In the following sections, we comprehensively discuss the approaches and scientific advancements of calcium-phosphate incorporation into zirconia scaffolds and zirconia incorporation into calcium-phosphate scaffolds, which have been investigated by numerous bone-tissue engineers.

Initiation of the application of CP-altered zirconia-based scaffolds for bioactivity, an enhancement was started in 2003. Kim et al., proposed a biocompatible HA coating (dip-coating) on the surface of a zirconia scaffold (synthesised via the sponge replication technique) [42]. In particular, to minimise the chemical conversion reaction between HA and zirconia, a fluorapatite (FA) layer was adopted as an intermediate. The authors found that the strength of the HA-coated zirconia scaffolds increased by a factor of 7 (compared with the HA), which is encouraging for load-bearing applications. Furthermore, the in vitro results confirmed that the surface-treated porous zirconia scaffolds significantly promoted the growth and proliferation of the osteoblast cells compared with untreated zirconia scaffolds. It is known that FA-inserted HA surface-treated zirconia scaffolds provide enriched bone-tissue regeneration properties. In 2004, Kim et al., varied the external coating with different calcium-phosphate coatings via the powder slurry method (TCP, HA, FA) or in the form of composites including TCP + HA and HA + FA on sponge replication technique-derived zirconia scaffolds [43]. Before the direct CP coating was formed, the zirconia scaffolds were coated with FA as an intermediate layer to minimise the interaction between the CP and the zirconia scaffolds. The authors found that the cell evolutions depended on the coating environment. For example, from the cell-differentiation output of MG63 cells, the alkaline phosphate establishment was found to be enriched in surface-treated scaffolds HA (HA, HA + FA and HA + TCP) compared with pure TCP- and FA-coated scaffolds. In 2004, the same group studied the effects of coating zirconia with HA via the sol-gel and slurry methods [44]. The dissolution rate in the case of the sol-gel-slurry coating was higher than that for the slurry coating. Additionally, the authors controlled the dissolution rate by adjusting the annealing temperature of the sol-gel HA layer. Osteoblast proliferation was confirmed by biological assessments using human osteoblast-like cells (MG63 cells).

After the successful formulation of zirconia scaffolds into strong and bioactive scaffolds via apatite dual-layers inner FA layer and an HA outer layer on the zirconia scaffolds, Kim et al., further authorised the clinical prospects of bioengineered zirconia scaffolds by performing in vivo studies on a rabbit calvarial defect model in 2008 [45]. To recognize the geometrical impacts on the bone-regeneration activities, the authors varied the porosity and pore size of the scaffolds. Remarkably, the bioengineered scaffold exhibited a higher porosity (~84–87%) and compressive strength (~7–8 MPa) than pure apatite-based scaffolds (~74% and ~2 MPa, respectively). Furthermore, according to the in vivo studies on the rabbit calvarial defect model, they proposed that the new-bone construction ensued effectively within the pore channels of all the apatite-engineered zirconia scaffolds, where the bone regeneration phases were similar to those of the pristine HA scaffold. Thus, Kim et al., confirmed the capability of apatite-engineered zirconia scaffolds for bone-regeneration applications by succeeding fundamental studies from in vitro to in vivo amendments, which is crucial for realising any academic research activities persevering further to real-world standards.

Likewise, in 2012, An et al., performed in vivo studies on a fabricated zirconia-HA composite [20]. The authors constructed zirconia/HA scaffolds via the conventional replication (polyurethane foam-scaffold) technique by dipping polyurethane sponge in the zirconia/HA slurries. The authors demonstrated that the compressive strength of the zirconia/HA scaffold increased from 2.5 to 13.8 MPa with an increase in the zirconia content from 50 to 100 wt%. Additionally, the authors found that the biological activity of the zirconia/HA scaffold was superior to that of pristine zirconia alone. More importantly, in vivo examination using fibrin gel comprising bone marrow-derived stromal cells (BMSCs) loaded with a zirconia/HA bioceramic scaffold offered a promising 3D surrounding for BMSC persistence and enriched the bone restoration near the implanted scaffold. The definite fabrication of the zirconia/HA scaffold via the polyurethane foam-scaffold technique and its successful bone regeneration in eight-week-old male SD rats are presented in Figure 1. Thus, An et al., demonstrated that bioinert zirconia can be used as an effective bone-generation material for larger bone defects with the aid of bioactive HA composite formation.

Interestingly, platelet-rich plasma (PRP) was introduced as a bone-growth-supporting agent to HA/zirconia scaffolds by Latifi et al., and Shahsavari-pour et al. [46,47]. In particular, zirconia scaffolds were produced via the polyurethane foam-replication technique and subjected to FA coating and HA coating via the slurry technique, followed by PRP/heparin sulfate (HS) impregnation to obtain HA/zirconia/PRP scaffolds. Scanning electron microscopy (SEM) of the HA/zirconia scaffolds revealed their high porosity, where the pores were impregnated with PRP gel, and the high-resolution SEM outputs of the PRP-impregnated HA/zirconia frameworks revealed nanoscale porosity (Figure 1b). Enriched osteoblastic proliferation and mineralisation of MG-63 cells were observed for the PRP/HS-impregnated scaffold. To examine the bone-restoration capability of the HA/zirconia scaffold with and without PRP, the authors performed an in vivo experiment, creating a rectangular bone defect in a rabbit mandible and replacing the defect using custom designed rectangular scaffolds (Figure 1c). The in vivo studies were conducted for a period of 8 weeks, and the HA/zirconia/PRP scaffolds were found to repair the artificial bone defects. The authors reported that the HA-zirconia scaffolds with and without PRP accurately imitated the bone mandibular properties in the short-term studies. Nonetheless, long-term observations revealed that PRP played no role in enhancing the synergic regenerative properties. It has been proposed that PRP has osteoinduction and antimicrobial activities.

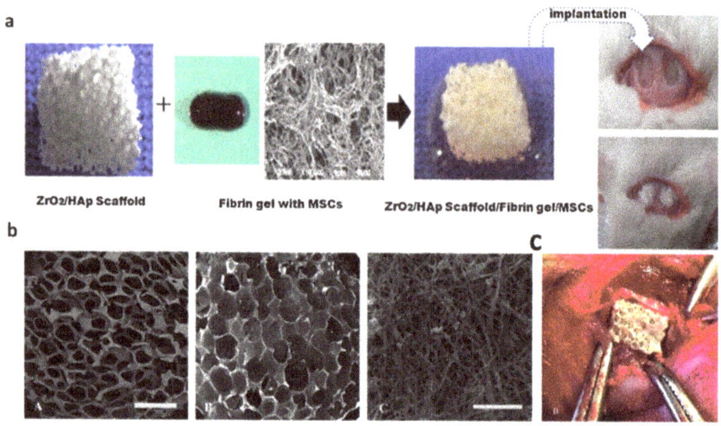

Figure 1. (**a**) Fabrication of the zirconia/HA scaffold via polyurethane foam-scaffold technique and its successful bone-regeneration in eight-week-old male Sprague dawley (SD) rats. Image is used from An et al., reprinted with permission from Elsevier [20], copyright© 2012. (**b**) PRP-impregnated HA/zirconia frameworks: (**A–C**) high-resolution SEM outputs of the PRP-impregnated HA/zirconia frameworks. (**c**) In vivo experiment for PRP-impregnated HA/zirconia frameworks; (**B**) in vivo demonstrations via a rectangular bone defect in the rabbit mandible and replacement of the defect using the deliberately designed rectangular scaffolds. Image used from Shahsavari-pour et al. [47], reprinted with permission from Elsevier, copyright© 2018.

Utilisation of β-TCP as a bioactive composite for tailoring the structural properties of zirconia scaffolds was proposed by Alizadeh et al., in 2014 [48]. In detail, the 3D β-TCP/zirconia (yttrium-stabilised) scaffold composite formation is favoured by external slurry mixing of β-TCP/zirconia at different wt% ratios (zirconia.Y_2O_3/β-TCP: A1:50/50, A2:40/60 and A3:30/70) and subsequent sponge formation using a polyurethane sponge (polyurethane foam-replica technique). The authors studied the physical and mechanical outputs with diverse β-TCP/zirconia ratios. They established that the porosity of the β-TCP/zirconia scaffolds can be changed from 65% to 85%. Subsequently, the authors confirmed that the compressive strength varied from 4.95 to 6.25 MPa with an increase in the zirconia content from 30% to 50%. The in vitro biological activity of the β-TCP/zirconia scaffolds was characterised using human endometrial stem cells, and it was found that the cell attachment and proliferation were enriched for the β-TCP/zirconia scaffolds with a ratio of zirconia.Y_2O_3/β-TCP 30/70. Conversely, Mohammad et al., reported that reinforcing the HA (75 wt%) matrix with 25 wt% zirconia enhanced the apatite-layer formation on the surface of the porous scaffold and increased the compressive strength to approximately 13.2 MPa [49].

A porous monolithic functionally gradient FG/BCP/zirconia scaffold with a cancellous bone structure was designed via the conventional polyurethane foam-replication technique by Lee et al. [50]. To recover the bioactive properties and reduce the amount of microdefects, the BCP/zirconia and BCP slurries were treated on the monolithic zirconia scaffold as an in-between layer and an external layer, respectively. The dimensions of the knitted pores and supports were approximately 100–250 μm and 110–300 μm after multilayer addition, which were appropriate for bone-renewal. The biological activity of FG/BCP/zirconia was evidenced by the rapid proliferation and cell attachment of the osteoblast-like MG-63 cells.

The use of dual bioceramic β-TCP/HA coatings on a polyurethane foam-replication technique-derived zirconia composite scaffold was described by Song et al. [51], and in vitro cellular-behaviour measurements provided evidence for the MC3T3-E1 pre-osteoblastic cell activity. Compared with titanium, the osteoblast activity on the surface-activated β-TCP/HA/zirconia composite scaffold was higher. Additionally, the osteoblast activity was mainly affected by the microstructure rather than the coating group. The short duration of the cellular-level investigation and the partial distribution of the coating materials in the interconnecting pores were limitations of this study that must be addressed in the future.

Furthermore, Lee et al., extended the BCP/zirconia scaffold study to design a unique multilayer BCl/zirconia scaffold with immobilised collagen surface modification (Col-BCP/zirconia) via a polyurethane foam-replication technique [52]. The average pore size of the scaffolds was in the range of 160–500 μm, which was sufficient for inducing new-bone growth [53]. In vitro cell proliferation and differentiation studies (MC3T3-E1 pre-osteoblast cell) revealed that the collagen-modified Col-BCP/zirconia scaffold was superior to the unmodified scaffolds. Collagen inclusion improved the cytocompatibility of the BCP/zirconia scaffold without affecting the bulk properties. More importantly, in vivo examinations of the Col-BCP/zirconia scaffold implanted into rabbit femurs after 1 and 5 months indicated that they have considerable promise for new-bone formation compared with the BCP/zirconia scaffold (Figure 2). Thus, the pioneering research outcomes highlight the importance of the Col-BCP/zirconia scaffold as an artificial bone material.

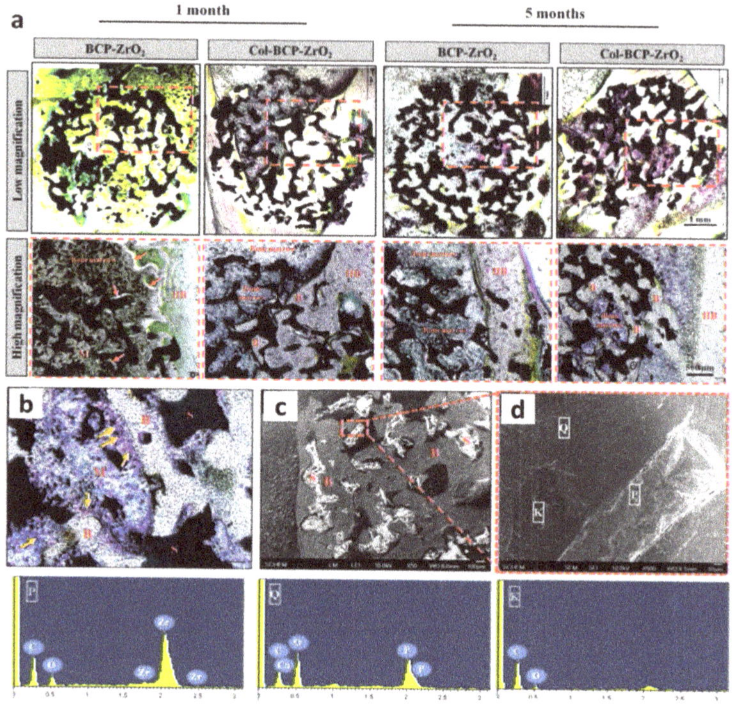

Figure 2. (**a**) Histological analysis of BCP/zirconia and Col-BCP/zirconia scaffolds 1 and 5 months after implantation in the femurs of rabbits, observed via Villanueva bone stains at low and high magnifications. HB: host bone; M: bone marrow; B: new-bone formation. (**b–d**) Histological analysis of Col-BCP-zirconia scaffolds after 5 months of implantation in the femurs of rabbits, observed via different methods. (**b**) Villanueva bone stain: HB, host bone; M, bone marrow; B, new-bone formation; S, scaffold. The SEM image in (**c**) is magnified in (**d**). The energy-dispersive X-ray spectroscopy profiles of P, Q and K were taken from (**c**). Images used from Lee et al. [52] reprinted with permission from Elsevier, copyright© 2015.

Additionally, 58S bioactive glass (BG58S)—a combination of calcium/phosphate units and silicate units—was reported to quickly bond with bone and stimulate bone formation, and it has been widely studied for dental implant applications [54]. This highly advantageous feature of BG58S was employed to modify the surface activity of zirconia scaffolds for the first time by Guimaraes et al., in 2019 [55]. Initially, zirconia scaffolds were designed using the conventional polyurethane foam-replication technique, with average pore diameters of 318, 423 and 564 μm, validating the 3D open-cell constructions. The BG58S coating was applied by immersing the zirconia scaffolds in a BG58S sol-gel solution. The coating thickness was controlled by optimising the viscosity of the sol-gel solution and altering the immersion rate, as shown in Figure 3a–f. The authors verified the in vitro biocompatibility of the zirconia scaffolds with and without the BG58S coating by employing MG-63 osteoblast-like cells (Figure 3g–i). The cell feasibility and proliferation were enhanced with a reduction in the pore size. Additionally, the scaffolds with the BG58S coating enhanced the cell viability and promoted cell proliferation, highlighting the importance of the chemical composition on the surface. The results of this study emphasise the significance of the chemical configuration on the surface, aperture diameter and microporosity in the utilisation of zirconia scaffolds as bone grafts.

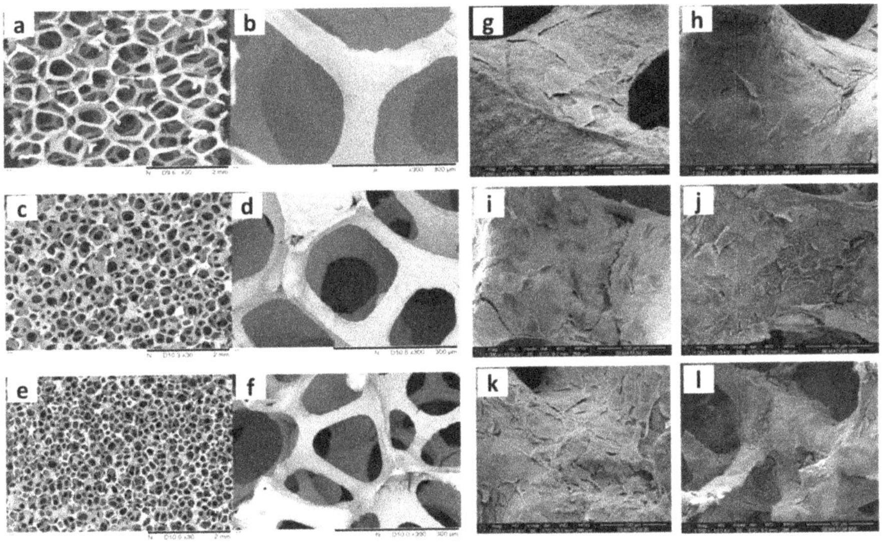

Figure 3. SEM images of 3D zirconia scaffolds (**a**,**b**) 564-Z, (**c**,**d**) 423-Z and (**e**,**f**) 318-Z at low (left) and high (right) magnifications. SEM images of MG-63 osteoblast-like cells on the uncoated 564-Z, 423-Z and 318-Z (**g**,**i**,**k**) and coated 564-Z.BG, 423-Z.BG and 318-Z.BG (**h**,**j**,**l**) scaffolds after 2 d of cultivation. Images used from Guimaraes et al. [55] reprinted with permission from Elsevier, copyright© 2019.

As mentioned previously, in 2020, Gouveia et al., used the protocol reported by Guimaraes et al., in 2019 for the synthesis of 58S bioactive glass and the fabrication of porous zirconia scaffolds reinforced with 58S bioactive glass [56]. Additionally, polyurethane sponges with 4, 60 and 80 pores per inch were fabricated. Moreover, the authors described an additional processing route that involved coating the pre-sintered zirconia scaffold template (1150 °C) with bioactive glass, followed by sintering at 1500 °C. The infiltration process occurred. Along with this processing step, an additional process was conducted, which involved BG58S coating and heat treatment at 600 °C.

Pristine zirconia grafts were studied for dental implant applications in the early 1980s because of their excellent aesthetic appearance and bioinert properties. However, in the available literature, there are limited research articles on the use of pristine zirconia grafts for bone-repair applications. Zhu et al., (2015) reported the optimisation of porous nano-sized zirconia scaffolds via a replica technique with a sufficient porous structure as that of a cancellous bone [57]. The porosity and pore size of the scaffold were controlled by changing the sintering conditions. The tremendous cell adhesion and enhanced proliferation of BMSC cells were observed after 14 d of incubation for a scaffold with 75.2% porosity. Additionally, the scaffold exhibited suitable mechanical properties for load-bearing applications.

Recently, Kim et al., (2018) studied the bone-formation abilities of pristine zirconia grafts engineered using the polyurethane-based sponge replication technique [27]. The designed grafts were subjected to in vivo studies to replace the bone defects in rabbit calvaria and compared with commercial graft materials including Osteon II (Os) and Tigran PTG (Ti). Even though the experimental groups achieved a great extent of new-bone development than compared with the defect group, the differences in the results among the experimental designs were insignificant owing to the similar granule sizes, shapes and porosities of the graft materials. Hence, it is important to design highly porous zirconia scaffolds with a definite bone size and shape using advanced techniques or to alter the surface of zirconia for enhancing the bioactivity.

Askari et al. [58] targeted the advancement of a computational framework (together with experimental confirmation) to regulate the mechanical characteristics of zirconia foams with different pore sizes (manufactured using the foam replica method) for bone-tissue reestablishment applications. Micro-computed tomography (CT) images were filtered to separate noise and smooth margins before fabricating 3D zirconia foams with an adaptive body-centred cubic background framework. The authors verified and scrutinised the stress distributions and magnitudes, scaffold deformation, stresses and plastic strains using the developed micro-CT-based finite-element model. The model was capable of depicting the mechanical stimuli on cells and the confined stress effort in the scaffolds.

2.2. Freeze-Drying Technique

Rather than fabricating zirconia in the form of a 3D structure, the 3D bioactive coating layer can be targeted over the zirconia surface. For example, the development of a thick, scaffold-like HA coating on durable zirconia substrates via a freeze-drying-assisted technique for fabricating porous scaffold-like HA/zirconia composites was proposed by Jiang et al. [59]. In vitro tests confirmed that the porous scaffold-like HA/zirconia composites were bioactive. Additionally, the 3D-HA coated onto the core zirconia retained sufficient mechanical properties for load bearing.

The foregoing studies mainly involved the use of calcium phosphate-based bioceramics with bone-like properties to reproduce the bone-graft activities of zirconia scaffolds. However, Teimouri et al., in 2015 established the use of inorganic polyoxometalates (POMs) with a unique biological property to promote the biological activity of zirconia by constructing a POM/zirconia/silk fibroin nanocomposite framework via the freeze-drying method [60]. The in vitro bioactive behaviour of the POM/zirconia/silk fibroin in simulated body fluid (SBF) was examined, and a uniform distribution of fibroblast cells on the POM/zirconia/silk fibroin composite scaffold was observed. The authors introduced a unique method for enhancing the bioactivity of bioinert zirconia; however, in vivo studies must be conducted to observe the effects of inorganic POMs in the human biological environment. The authors performed another study based on silk fibrin (SF) and nano-zirconia; however, the POM polymer was replaced by chitosan (2%) to fabricate a biocomposite scaffold with an interconnected porous structure using a freeze-drying technique [61]. The compression strength and modulus of the composite scaffold were twofold higher than those of the polymer scaffold (SF/CS) owing to the existence of zirconia in the polymer matrix. Moreover, in an evaluation of human gingival fibroblast (HGF) cells, the composite scaffold exhibited higher biocompatibility.

Recently, a comparative investigation of the physical and biological properties of chitosan-nano-HA (CS-nHA), chitosan-nano-zirconia (CS-nZrO) and chitosan-nano-calcium zirconate (CS-nCZ) porous composite scaffolds (Figure 4a) fabricated via freeze-drying for bone-reestablishment applications was performed by Gaihre et al. [62]. The in vitro activity of the OB-6 pre-osteoblast cells was superior to that of the extended filopodia on CS-nHA and CS-nZrO compared with the CS-nZrO composite scaffolds (Figure 4). The authors expected to investigate further studies on the osteogenic capability of CS-nZrO composite scaffolds.

2.3. Pore Former/Space Holder Technique

The vacuum slip casting technique was used to construct an alumina/zirconia composite scaffold with the aid of expanded polystyrene (EPS) beads (acting as pore formers) by Liu et al. [63]. The designed scaffold exhibited homogeneously circulated interconnected pores. The alumina/zirconia composite scaffold was further coated with a thick bioactive glass (58S33C) layer. The bioactive glass-coated alumina/zirconia scaffolds exhibited optimal porosities (60–66%), high strength (5.42–7.52 MPa) and enhanced bioactivity (apatite-layer formation in the SBF solution after 24 h). The authors showcased multiple bioactive glass-coated macropores as permanent scaffolds for bone-tissue restoration.

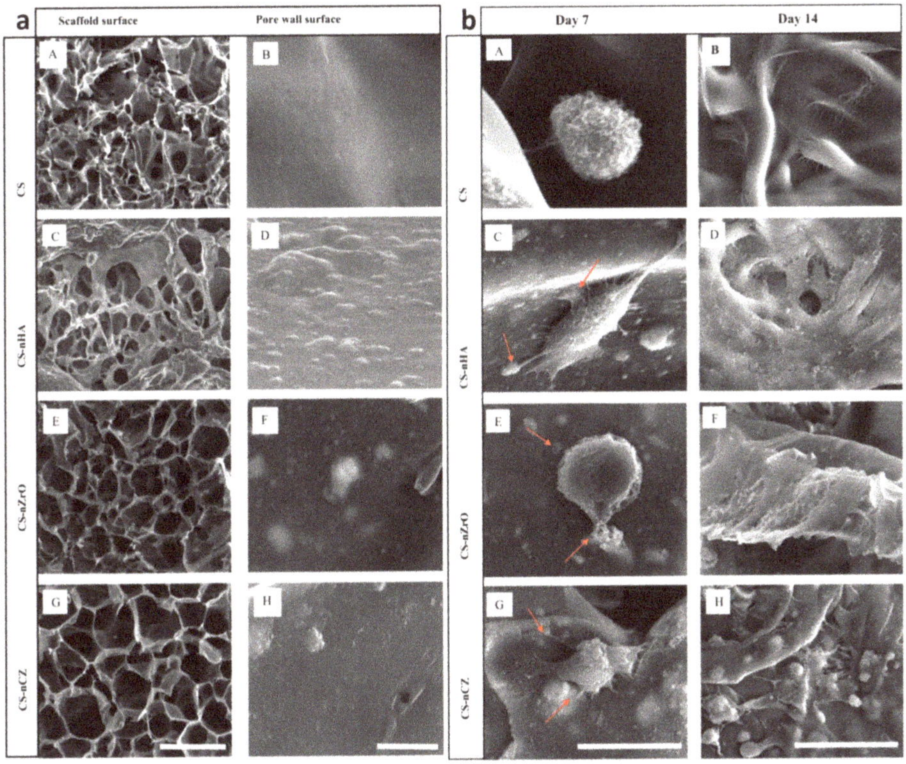

Figure 4. (**a**) SEM images showing the highly porous morphology of the scaffold surface (**A**,**C**,**E**,**G**) and magnified images showing the surface of the scaffold wall (**B**,**D**,**F**,**H**); the scale bars indicate 100 and 2 μm for the scaffold surface and the pore wall surface, respectively. (**b**) SEM images (**A**–**H**) showing the morphology of pre-osteoblast attached and proliferating along the surfaces of different scaffolds; the scale bars indicate 20 and 50 μm in the images taken on days 7 and 14, respectively. Images used from Gaihre et al. [62] reprinted with permission from Elsevier, copyright© 2019.

In one of the reported studies, the bone-growth activities of yttria-stabilised zirconia (YSZ) and magnesium-stabilised zirconia (MgSZ) porous scaffolds were compared (the scaffolds were formed using EPS beads) [64]. The MC3T3-E1 pre-osteoblast cell activity on the YSZ was enriched owing to the surface chemical dissimilarities between the YSZ and the MgSZ. However, the authors claimed that both YSZ and MgSZ porous ceramics can be effective scaffolding materials for cortical bone, because of their comparable mechanical and biological properties.

2.4. Solvent Casting and Salt Leaching

Solvent casting and particulate leaching are among the simple and time-effective techniques for producing bioceramic scaffolds with the desired porosity [65]. Unique PLA-HA-YSZ nanocomposite scaffolds with diverse compositions were fabricated via solvent casting and particulate leaching by Ziaee et al. [66]. Among all the ratios tested, a PLA15%—HA-15%YSZ nanocomposite scaffold exhibited the highest compressive strength and SBF activity. The compressive strengths of the scaffolds were reduced after they were soaked in SBF, and the scaffolds containing HA underwent larger strength reductions than those containing YSZ.

In 2018, Mokhtar et al., conducted a case study to assess the efficacy of a custom-made porous zirconia scaffold along with a buccal trapezoidal flap for closure of oroantral fistula [67]. This study was performed on 10 patients suffering from oroantral defects due to extraction of the first and second premolars in a maxillary region with dimensions of approximately 6–9 mm. Initially, a virtual bone model with defects was constructed using the stereolithography photopolymer-based 3D-printing technique from cone-beam computed tomography (CBCT). The zirconia scaffold was prepared via solvent casting and the salt-leaching technique, and the fitness was assessed with the bone model before the sterilisation process. Clinical (extraoral and intraoral examinations) and radiological (CBCT) assessments were performed before and after surgery. Postoperative clinical follow-up evaluations were performed at 2 weeks, 1 month and 3 months. A radiographic examination was conducted after 2 weeks using panoramic radiography. CBCT was performed after 3 months to evaluate the bone formation; the bone density was compared with that during preoperative CBCT. The intensity of pain and frequency of minor complications (postoperative bleeding/edema) were significantly reduced over time, while the bone density increased by approximately 41.2%. The authors found that the zirconia scaffolds with interconnected porous structures enrich the formation of new bone in oroantral defects.

3. Fabrication of 3D Zirconia-Based Scaffolds via the Digital Technique

3.1. Computer-Aided Design/Computer-Aided Milling (CAD/CAM) Technique

After the acclaimed applications of zirconia ceramics in dental crowns and implants, zirconia ceramics were expected to dominate the biomedical field. However, the direct utilisation of zirconia scaffolds to reconstruct bone is also initiated in the maxillofacial reconstruction area in the dental field. Aftan et al., developed an innovative method for mandibular reconstruction [68]. The authors designed a patient-specific zirconia prosthesis via the CAD/CAM technique using a zirconia block. They employed the zirconia prosthesis technique for the restoration of mandibular flaws in a human trial. This case study was performed on 20 patients for 62 months, with defects caused by mandibular trauma, tumours, and congenital abnormalities, which required surgical resection and reconstruction. A few patients underwent two surgeries; the Boweman Conroyd appliance was placed in the first surgery, and it was replaced by the predesigned zirconia prosthesis in the second surgery. The zirconia prosthesis was designed according to the patient defects and surrounding anatomical structures using a 3D CAD model. Milling was performed on the zirconia block, followed by sterilisation. For the placement of a dental implant, the authors voluntarily created a hole in the prosthesis. The results were encouraging, with a 95% success rate, and the reconstruction did not affect the mandibular function or aesthetics. Minor complications (pain and edema) were reported.

It is well known that CAD/CAM is a destructive technique that cannot produce zirconia scaffolds with high porosity. Recently, an interesting approach for designing zirconia scaffolds with ~40% porosity and a multilayer assembly via CAD/CAM was proposed by Marques et al. [69]. The distinctive design strategy involves 5-axis milling (XYZ axes and rotating axes (A and B axes)) that allows the part to move 360° in both directions and 20° towards the front and back [69]. A schematic of the zirconia scaffolds designed using the CAD/CAM model is presented in Figure 5a. The scheme comprehensively illustrates the process of designing complex zirconia scaffolds using a modified CAD/CAM model, followed by mechanical testing (compression testing), whereby the Young's modulus values were found to be associated with the host bone. The rapid diffusion behaviour of water inside the channels of the zirconia scaffolds was established using capillary testing. Likewise, to replicate the endosseous implant, the authors demonstrated implant insertion imitation practice, wherein the implant exhibited superior fixation at the early stage of implantation. Tribological tests revealed that the cavities/valleys inside the scaffolds were loaded by bone. Even though the authors mimicked the bone implantation of the zirconia scaffolds via tribological tests, it is essential to verify the cell activity of the scaffolds in the biological environment.

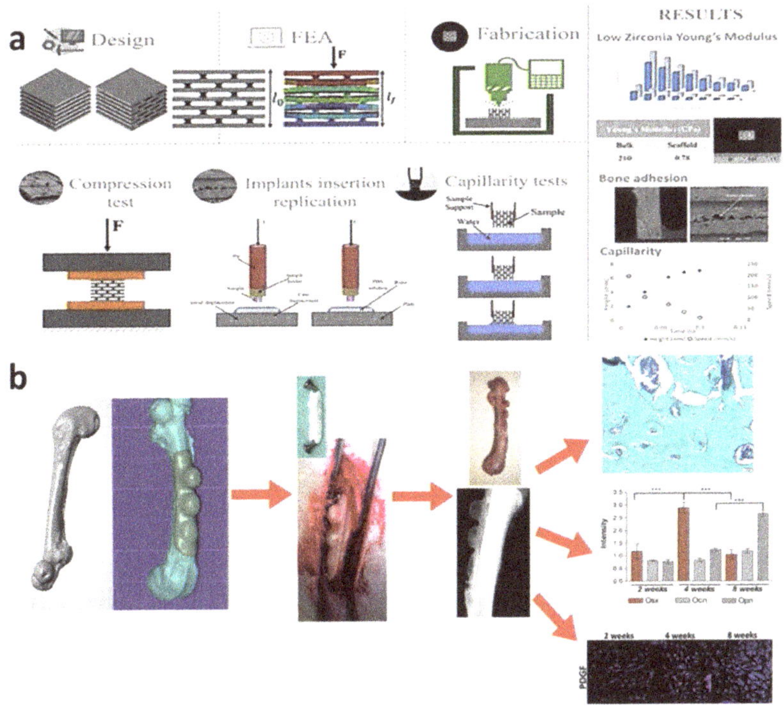

Figure 5. (**a**) Schematic of zirconia scaffolds designed via a CAD/CAM model. Image used from Marques et al. [69] reprinted with permission from Elsevier, copyright© 2019. (**b**) Detailed illustration of the design and milling of 3D-tailored CAD/CAM zirconia space-maintaining strategies for a rat femur. Image used from Tetre et al. [70] reprinted with permission from Elsevier, copyright© 2021.

To all the above, most recently, Tetre et al., revealed a unique method for evaluating bone reconstruction using zirconia scaffolds in the rat femur via a 3D-CAD/CAM approach [70]. A thorough illustration of the design and milling of 3D-tailored CAD/CAM zirconia space-maintaining strategies for a rat femur is presented in Figure 5b. As shown, a 3D-customised assembly was produced with different heights. The authors verified the guided bone regeneration (GBR) abilities of the rat femur via Gomori's trichrome histomorphometrical examination. The authors monitored GBR activation at different time intervals (2, 4 and 8 weeks) and found that the Haversian system was present in freshly grown bone. Thus, the tailored milled zirconia scaffold in this study provided a detailed understanding of progressive bone-tissue construction. The successful bone-reconstruction ability of the zirconia scaffold highlights the effectiveness of the 3D-tailored CAD/CAM technique for extending its application to various orthopaedic applications.

3.2. Extrusion-Based Techniques

3.2.1. Multi-Pass Extrusion Technique

Following the clinical success of an extrusion-based technique to fabricate zirconia scaffolds for dental applications, the design of 3D zirconia scaffolds for bone-regeneration applications has been widely studied [23]. Lee et al., analysed this technology even before the worldwide upsurge in zirconia scaffolds for dental applications [71]. They fabricated HA-coated micro-channelled fibrous Al_2O_3-(monoclinic, M-zirconia)/(tetragonal, t-zirconia) composite scaffolds via the multi-extrusion process in 2004 [72]. In a preliminary study, they developed a multi-extrusion printing technique for designing zirconia composite scaffolds with good mechanical properties. In another study, the research group varied several printing parameters, including the pore-gradient rate, extrusion ratio and microstructure, to optimise the experimental conditions for the fabrication of Al_2O_3-zirconia composite scaffolds [73]. Their fabrication technique involved the mixing of bioceramics using the polymer ethylene vinyl acetate, followed by extrusion. The extrusion process is repeated to build a constantly porous core assembly with alternating Al_2O_3 and zirconia layers. Using this custom design, the authors constructed various scaffolds (both individual scaffolds and binary mixtures) over the subsequent years to achieve unique interconnected architectures [71,74,75]. Later, in 2006, they demonstrated the in vitro understanding of Al_2O_3-(M-zirconia)/t-zirconia composite scaffolds with well-established 3D-interconnected micropores using human osteoblast-like MG-63 cells. The MG-63 cells were attached on the rough outer layer as well on the micropores inside the scaffold body [75]. In the following years, owing to the repeated efforts of Lee et al., various structural advantages have been implemented, yielding diverse zirconia composite scaffolds that are useful for bone-regeneration applications. For example, using the multi-extrusion process, Lee et al., constructed bone-like continuously porous TCP/TCP-(t-zirconia)/t-zirconia composites (originally, they used HA that transformed into TCP during heat treatment) (Figure 6a) [76]. To rectify the microcracks upon sintering due to thermal expansion mismatch issues, the authors made the boundary amongst the pores with three types of laminates (Figure 6b–d). Cross-sectional SEM study confirmed the uniform pore channels and the absence of delamination and cracks after the sintering process. The Haversian channel dimension and outer cortical sample size were linked at diverse sintering temperatures, and their values were recognised to be 86 μm, 10.3 mm and 53 MPa, respectively. Furthermore, complete adhesion and distribution of osteoblast-like MG-63 cells were observed in in vitro survival studies of the artificial cortical bone developed by the authors (Figure 6e–j).

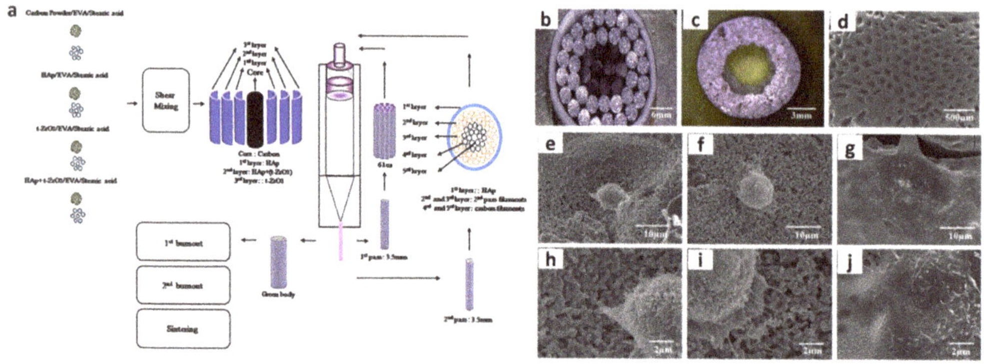

Figure 6. (**a**) Schematic of the multi-pass extrusion process. Photographs taken before extrusion and after sintering (**b**) arranged filaments and shell before final (3rd) extrusion, (**c**) sintered bodies at 1500 °C, (**d**) enlarged SEM image of the pores. SEM images of MG-63 cell morphologies on TCP/TCP-(t-zirconia)/t-zirconia composites at incubation times of 15 min (**e**,**h**), 30 min (**f**,**i**) and 60 min (**g**,**j**). Images used from Lee et al. [76] reprinted with permission from Elsevier, copyright© 2011.

The combined mechanical and biological outputs of the TCP/TCP-(t-zirconia)/t-zirconia scaffolds in this study were proven to be an ideal composite favourable for synthetic cortical bones such as toe joints and finger replacements. This was followed by the authentic demonstration of TCP/TCP-(t-zirconia)/t-zirconia scaffolds as an artificial cortical bone. Lee et al., constructed porous multilayer HA/t-zirconia scaffolds via a multi-extrusion process comprising concentric laminated structures and micro-channelled groups [77]. Microstructural studies revealed that the HA/t-zirconia scaffold consisted of alternating units of HA, HA/t-zirconia and t-zirconia. The mechanical properties of the scaffolds varied significantly with the increasing temperature. The scaffold sintered at 1400 °C exhibited a high compressive output (20 MPa), and no cracks were observed, whereas the sample sintered at 1500 °C exhibited a noticeable crack (Figure 7a–f).

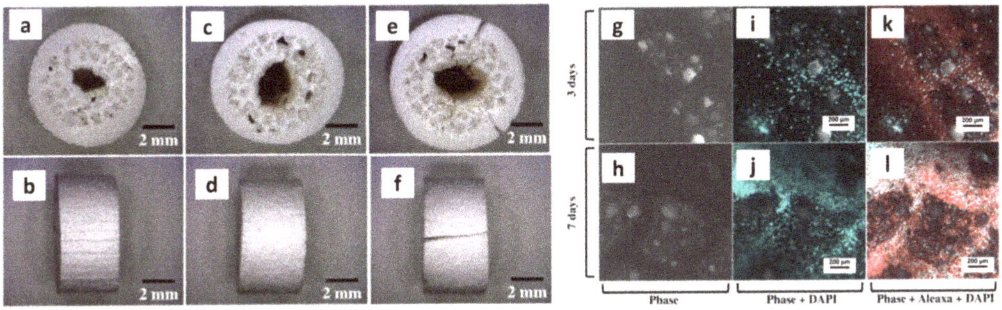

Figure 7. Photographs of HAp/t-zirconia composites sintered at 1300 °C (**a**,**b**), 1400 °C (**c**,**d**) and 1500 °C (**e**,**f**). Confocal microscopy images of osteoblast cell proliferation on an HAp/t-zirconia composite scaffold sintered at 1400 °C for 3 days (**g**,**i**,**k**) and 7 days (**h**,**j**,**l**). Reprinted with permissionfrom Lee et al. [77], copyright© 2012.

The Haversian passage and interior dimensions were calculated to be 80 μm and 1.9 mm, which were favourable for cell development. Furthermore, in vitro osteogenesis was confirmed by the uniform proliferation of osteoblast-like MG-63 cells (Figure 7g–l). Additionally, the multi-extrusion methodology was employed by Lee et al., to construct a macroporous channelled scaffold comprising a thin zirconia covered with a sacrificial background layer made of PCL/BCP, which undergoes degradation to promote new-bone formation [78]. The zirconia-PCL/BCP multilayer scaffold with the unidirectional cylindrical channels exhibited 78% porosity, along with an acceptable compressive strength (12.7 MPa). The in vitro proliferation of osteoblast-like MG-63 cells inside the pore wall of the zirconia-PCL/BCP multilayer scaffold was confirmed by SEM. Unfortunately, there is not much acknowledgement of Lee et al.'s advancements. We take this opportunity to provide a comprehensive summary of their research results. This will provide readers with a better understanding of the evolution and progress of 3D zirconia scaffolds, promoting their use as standard materials for bone-tissue engineering.

3.2.2. 3D-Bioplotter Technique

Sapkal et al., used the extrusion-based technique and 3D-Bioplotting to construct 3D β-TCP/zirconia composite scaffolds with diverse architectures for healing large bone deficiencies [79]. Scaffolds (0°–90°, 0°–45°–135°–180°, 0°–108°–216° and 0°–72°–144–36°–108°) tactically designed by the authors with diverse filament alignments (Figure 8a–h) were systematically subjected to mechanical and biological testing evolutions. The in vitro bioactivity capability of the β-TCP/zirconia composite scaffold to replicate the human-bone environment was assessed using MG-63 human osteosarcoma cells. Among the samples tested, the β-TCP/zirconia composite scaffold with the 0°–72°–144°–36°–108° filament orientation exhibited the best mechanical strength and cell-breeding capability.

Moreover, they concluded that the β-TCP/zirconia composite with a 70/30 ratio is favourable for large-size bone-repair applications. The authors proposed future research directions to extend their research to the next level, i.e., the insertion of Haversian canals inside the β-TCP/zirconia composite scaffolds, for imitating blood vessels and stimulating vascularisation.

Fu et al., highlighted the importance of zirconia reinforcement in β-Ca_2SiO_4 scaffolds for bone-tissue engineering constructed via the 3D printing of silicone resin-loaded $CaSiO_3$/zirconia fillers using the extrusion-based 4th Bioplotter [80]. The amount of zirconia incorporated into $CaSiO_3$ was varied from 0 to 15 mol%, and the scaffolds were denoted as (0Zr-C2S, 5Zr-C2S, 10Zr-C2S and 15Zr-C2S). The β-Ca_2SiO_4/zirconia scaffolds designed in this study exhibited a rough surface and micropores, as indicated by the SEM results in Figure 9a–p, which are important features for mimicking bone behaviour. The engineered β-Ca_2SiO_4/zirconia scaffolds with a woven macropore network exhibited a high porosity (~67%) and compressive strength (~6.1 MPa) favourable for load-bearing bone reconstruction. The enhanced biological activity of the 15Zr-C2S scaffolds in the in vitro rat bone marrow mesenchymal cells encouraged the authors to examine the new-bone formation capability in a real biological environment (in vivo studies). A histological examination revealed that the bone restoration was more pronounced in 15Zr-C2S scaffolds than in a pure β-Ca_2SiO_4 scaffold implanted in rat deficiency zones (Figure 9q,r). In summary, zirconia incorporation induces dual properties to β-Ca_2SiO_4 scaffolds (i.e., improved mechanical strength and controlled degradation of Ca and Si), which is essential for renewed bone growth.

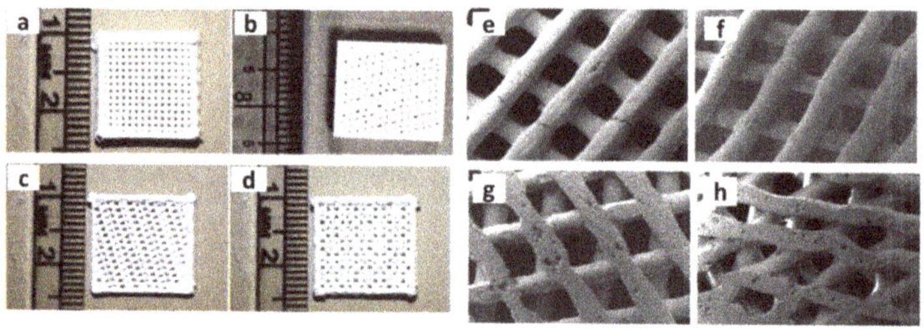

Figure 8. Photographs (**a–d**) and SEM images (**e–h**) of Sintered β-TCP/zirconia scaffolds with different laydown patterns and (**a**,**e**) 0°–90°, (**b**,**f**) 0°–72°–144°–36°–108°, (**c**,**g**) 0°–108°–216° and (**d**,**h**) 0°–45°–90°–135°, respectively. Reprinted with permission Images used from Sapkal et al. [79], copyright© 2017.

3.2.3. Fused Deposition Modelling (FDM) Technique

Sa et al., fabricated reinforced BCP scaffolds with different amounts of zirconia (BCP/zirconia) in 2018 via the FDM 3D-printing technique [81]. A photograph of the FDM process employed by Sa et al., to design zirconia-reinforced BCP scaffolds. SEM of the BCP/zirconia scaffold with overall dimensions of 6.0 × 6.0 × 3.0 mm^3 revealed that the pore size was 350 μm (Figure 10). Incorporation of 10% wt% zirconia powder into the BCP scaffolds enhanced the compressive strength without degrading the bioactivity. The authors verified the in vitro bioactivity of the BCP and BCP/zirconia scaffolds in the presence of human mesenchymal stem cells (hMSCs) by comparing the fluorescence staining of the hMSCs with BCP and BCP/zirconia scaffolds. The BCP/zirconia scaffolds exhibited improved expression of bone morphogenic protein-2 compared with the pristine BCP scaffolds. The synergistic effects of zirconia-stimulated scaffolds in achieving an active culture upon osteogenic differentiation of bone tissues were proposed by Sa et al.

In addition to the traditional bioceramics, biopolymers have been investigated for bone-regeneration applications because of their easy processability and controlled degradation amount.

Polycaprolactone (PCL) is the most widely studied biopolymer because of its poor mechanical properties and poor hydrophilicity. PCL alone cannot be used directly as a bone-tissue framework. Hence, similar to most of the bioceramic designs, researchers have attempted to include zirconia to enhance the mechanical properties of PCL [82]. For instance, Wang et al. [83] embedded nano-zirconia with different weight ratios (5, 10 and 20 wt%, i.e., PZ5, PZ10 and PZ20, respectively) into PCL (PCL/zirconia) through the melt mixing method and fabricated ordered scaffolds using the extrusion-based 3D-printing technique. They have found that the pristine PCL, the PCL/zirconia scaffolds exhibited improvements of 50% and 40% in the compressive strength and Young's modulus, respectively. Furthermore, the improved hydrophilicity and better water-adsorption ability of the PCL/zirconia scaffold are conducive to cell bonding and nutrient passage. For instance, an in vitro bioactivity survival test of PCL/zirconia scaffolds using the MC3T3 cells via laser confocal microscopy revealed that the cell growth rates of the PCL and composite scaffolds (PZ5, PZ10 and PZ20) at 1, 3 and 7 d were >90%. The PZ20 scaffold exhibited superior MC3T3 cell distribution, and the degree of cell proliferation varied with respect to the zirconia content. Thus, the strategic design of PCL/zirconia scaffolds proposed by Wang et al., confirmed the effectiveness of zirconia incorporation in transforming the PCL into a potentially feasible scaffold for bone-tissue restoration.

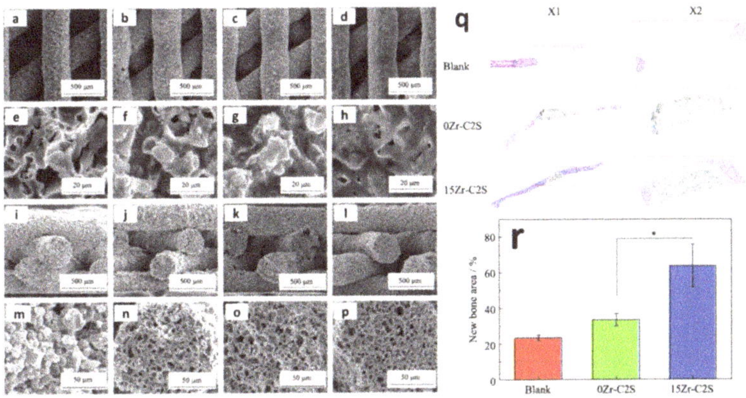

Figure 9. (**a–p**) SEM images of (**a**) 0Zr-C2S, (**b**) 5Zr-C2S, (**c**) 10Zr-C2S and (**d**) 15Zr-C2S scaffolds, with SEM fracture surface images from (**i,m**) to (**i–p**). In vivo results for 0Zr-C2S and 15Zr-C2S scaffolds (**q**) histological and (**r**) quantitative analysis results for newly formed bone. Reprinted with permission from Fu et al. [80], copyright© 2019.

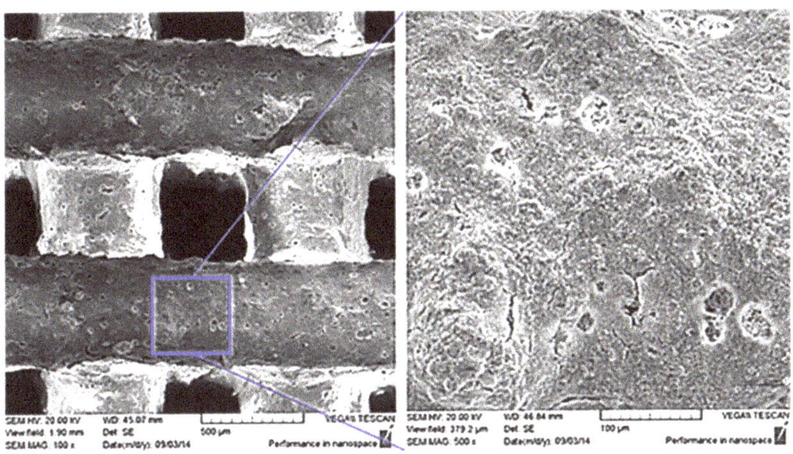

Figure 10. SEM images of the BCP/zirconia scaffold. The left image shows a pore size of approximately 350 μm and printed lattice width of 500 μm. The right side depicts a granular surface of the 3D-printed scaffold after printing and sintering. Images used from Sa et al. [81] reprinted with permission from Willey and sons, copyright© 2018.

3.2.4. Robocasting/Direct Ink Writing (DIW) Technique

Robocasting, i.e., DIW, techniques allow the design of high-quality complex ceramic structures via optimal formulation and deposition of colloidal ceramic ink/paste that flows through the printing nozzle with a high solid loading content to form a 3D construct with open porosity and are widely used owing to their cost-effectiveness and easy processing [84].

Few attempts to fabricate zirconia scaffolds using robocasting for replacing bone defects have been reported. The formulation of a zirconia suspension with high solid loading and the optimal rheology plays a crucial role in printing; thus, Brazete et al., investigated the dispersibility of zirconia powder (45 vol%) in a homogeneous aqueous suspension with different ratios (0.3%, 0.5%, 0.8%) of anionic dispersant (Dolapix CE 64) [85]. With 0.3 wt% dispersant, the suspension exhibited shear-thinning rheological behaviour. Furthermore, the authors deliberated the optimised 0.3 wt% dispersed ink with higher zirconia loading of 48 vol% for robocasting.

In 2015, Li et al., reported the fabrication of 3D cylindrical and woodpile-like 3D zirconia scaffolds with interconnected scaffolds via the DIW technique [23]. The DIW design proposed by the authors comprises three major parts: the computer-aided system (CAD), extrusion needle and X-Y-Z platform. They unlocked the access to design 3D zirconia tissue engineering scaffolds with water-based zirconia ink having a zirconia content of 70 wt%. The zirconia ink was printed into layer-by-layer ordered 3D zirconia scaffolds. The compressive strength of the 3D zirconia scaffolds (10 MPa) was higher than that of the HA scaffolds. The in vitro proliferation of HCT116 cells on the 3D zirconia scaffolds was verified using a microscope. Thus, the authors successfully proposed the possibility of the DIW method to design scaffolds with precise control of the porosity for advanced bone-tissue engineering applications.

Recently, Kocylo et al. [86] engineered 3D zirconia scaffolds with 61 and 75.3 vol% porosity via the direct ink-writing method. The SEM results in Figure 11 confirm the uniformity of the zirconia scaffolds with good control in thread and pore openings. With changes in the porosity level, the compressive strength of the printed zirconia scaffolds varied between 20.8 and 62.9 MPa. The authors applied multi-surface alteration tactics involving a dual apatite FA/HA coating to the 3D zirconia scaffolds for obtaining zirconia/FA/HA composite scaffolds (originally proposed by Kim et al. long ago for 3D zirconia scaffolds). For comparison, they applied the HA/HA coating to the 3D zirconia (zirconia/FA/HA scaffolds). The authors performed an in vitro investigation to understand the bioactivity and dissolution properties of the zirconia/FA/HA and zirconia/FA/HA composite scaffolds, which were immersed in SBF and a physiological saline solution (0.9%, NaCl) for 28 d. The zirconia scaffolds with the FA/HA coating exhibited enriched bioactivity and slower dissolution than the zirconia scaffolds with the HA/HA coating. In contrast to other reported studies, Stanciuc et al., developed a new platform with the processing, structural characterisation and interaction of human primary osteoblast cells of a robocast ZTA scaffold [87].

3.3. *Photopolymerisation-Based Techniques*

Since the introduction of photopolymerisation (PP) based 3D printing by Hull in 1986 [88], it has been claimed that PP-based practices have considerable potential for producing 3D bioceramic scaffolds with high accuracy [89].

3.3.1. Digital Light Processing (DLP)

DLP technology has been widely studied for the fabrication of ceramic parts, as it can produce complex ceramic structures with a high precision, fast processing and a high accuracy compared with other 3D-printing technologies. The slurry suspension was prepared by optimising the solid loading of the ceramic, organic monomer, dispersant and photoinitiator, and the slurry was irradiated with ultraviolet light. The rheological property of the ceramic suspension is a key factor in the 3D-printing process and affects the quality of the final product. However, the use of DLP technology in the biomedical field has been restricted owing to the incorporation of the photosensitive resin and photoinitiator which is known to be cytotoxic, into the slurry. Therefore, successive de-binding and sintering processes must be developed before this scaffold can be employed for bone-tissue regeneration [90].

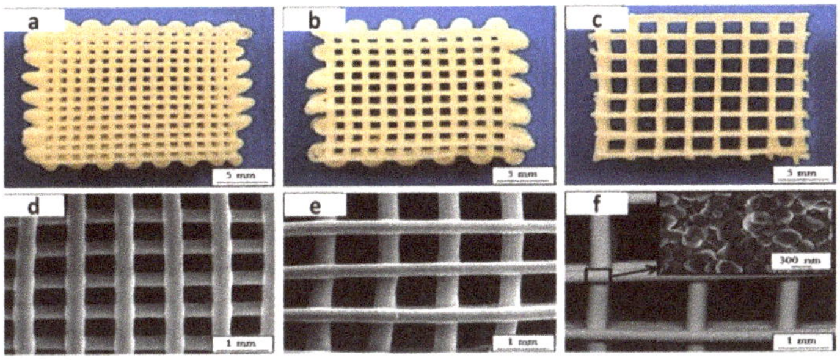

Figure 11. Optical images of 3D periodic zirconia scaffolds with different minimum separation distances between rods (**a**) 0.7 mm, (**b**) 0.9 mm and (**c**) 1.8 mm and corresponding top-surface images (**d**–**f**), respectively. The inset SEM image shows the high degree of sintering achieved within the filaments. Image used from Kocylo et al. [86] reprinted with permission from Elsevier, copyright© 2021.

Few in vitro investigations of zirconia-based scaffolds fabricated using the DLP technique have been reported. In 2020, Cao et al., evaluated the biological activity and mechanical strength of a zirconia/HA porous scaffold fabricated via DLP (Figure 12) [91]. They added HA at various concentrations (0%, 10%, 20% and 30%) to zirconia to enhance the bioactivity of the porous scaffold with partial degradation. With the increasing HA content, the degrees of cell proliferation and differentiation of MC3T3-E1 cells increased, while the compressive strength decreased. However, HA10% exhibited improved strength compared with the control (zirconia) scaffold.

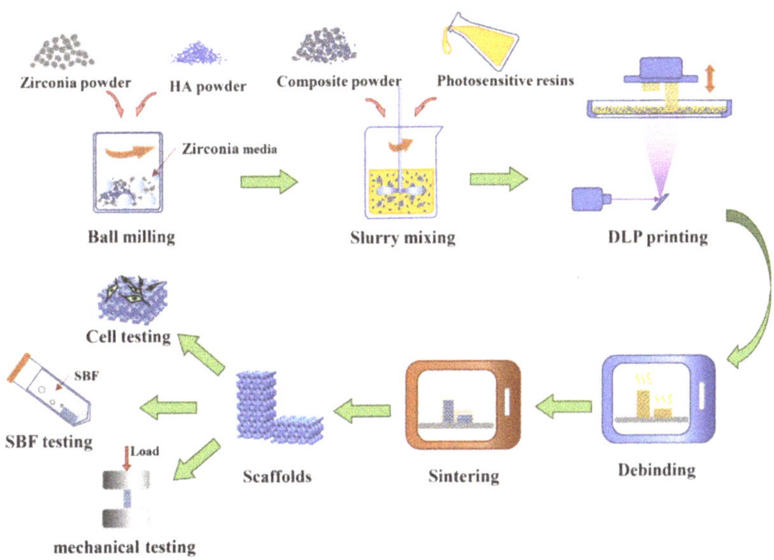

Figure 12. Illustration of the fabrication and testing process. Image used from Cao et al. [91], reprintedwith permission from Elsevier, copyright© 2020.

In contrast, Zhang et al., fabricated a porous zirconia toughened HA composite scaffold for bone-tissue engineering using the DLP 3D-printing process with HA as the primary ratio of the scaffold; additionally, a small amount of zirconia was used as a structural stabiliser [26]. The ceramic powder (60 wt%) with a high solid loading into the suspension by modifying the surface of the HA nanoparticles with two organic modifiers (KH570 and oleic acid) before adding to the acrylic resin and for improved dispersion was proposed. Castor oil phosphate was added to reduce the viscosity of the ceramic suspension. The authors reported that the incorporation of zirconia into HA reduced the decomposition of the HA phase and promoted densification during the firing process. The de-binding and sintering of the printed scaffold were performed in a vacuum high-temperature furnace to avoid the generation of internal cracks and defects in the sintered ceramic parts, which are caused by the internal pressure created inside the samples during sintering. The mBMSC exhibited no cytotoxicity in the 3 wt% zirconia group and exhibited increased cell proliferation over a period of time.

3.3.2. Selective Laser Sintering (SLS) Technique

SLS is a well-established additive-manufacturing technique that employs laser energy as a primary source to fuse powder particles into the desired 3D architecture [92]. It has been widely studied in the field of biomedical applications for the fabrication of 3D structures. Shuai et al., for the first time attempted to fabricate a nano-zirconia-reinforced calcium silicate ($CaSiO_3$) porous scaffold ($CaSiO_3$/nano-zirconia scaffold) via the SLS technique for bone-recreation applications with improved mechanical properties [93]. The authors found that increasing the amount of nano-zirconia resulted in superior mechanical properties. However, when the zirconia concentration exceeded 30%, the zirconia affected the sintering, caused unsolicited agglomeration and degraded the materialising ability. An in vitro analysis revealed that the $CaSiO_3$/nano-zirconia (20%) scaffolds had the ability to mock bone features by forming a bone-like apatite layer and promoting superior assembly of MG-63 cells. The authors concluded that the porous scaffolds with the addition of 20% of nano-zirconia exhibited good biocompatibility and mechanical properties.

3.4. *Electrospinning*

Electrospinning is a unique method in which an electric field induced by a high direct-current voltage between a roller connection and a needle tip is used to spin bioceramic fibers with a constant diameter into an interconnected flexible bioceramic scaffold [94–96]. With the increasing demand for scaffolds with nano-to-microscale simulation of the configurations of the extracellular matrix (ECM), in 2016, Gazquez et al., proposed the fabrication of YSZ nanofiber oriented scaffolds for bone-tissue restoration via electrospinning a mixture of YSZ and polyvinylpyrrolidone [97]. In contrast to rigid bone ceramics, the designed YSZ scaffolds have excellent flexibility at the macroscale (Figure 13a). In an in vitro bioactivity survival test using hMSCs outputs, the nanofibrous YSZ scaffolds supported the attachment and proliferation of the seeded mesenchymal stromal cells, as confirmed by SEM and hMSC phalloidin staining analyses (Figure 13b–g). The authors proposed an alternative approach for designing flexible zirconia scaffolds to traditional fragile ceramics for bone-tissue engineering. In a recent study, Thakare et al., focused on the engineering of PCL/zirconia composite nanofiber scaffolds by varying the amount of zirconia from 6% to 30% via the electrospinning process [98].

The interwoven PCL/zirconia composite nanofiber scaffolds fabricated by the author had high material stability and porosity, which are favourable for imitating the dynamic atmosphere of natural tissues. Compared with the pristine PCL scaffold, PCL/zirconia composite nanofiber scaffolds exhibited controlled degradation, swelling and superior bioactivity. Additionally, the PCL/zirconia composite nanofiber scaffolds were nontoxic, as indicated by a cell-viability test using the 3T3 mouse fibroblast cell line. This study further confirmed that electrospinning can be used to fabricate PCL/zirconia composite nanofiber scaffolds that are favourable for tissue-engineering.

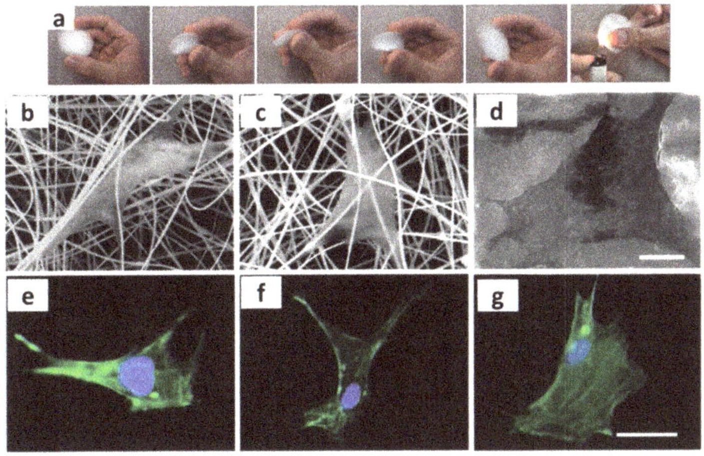

Figure 13. (**a**) Flexibility of YSZ nanofiber mats. Characterisation of the cell morphology and actin cytoskeleton organisation of hMSCs after 3 d of culturing in BM. Representative SEM images of hMSCs on (**b**) microwave-annealed (MW 5 °C/min) nanofiber scaffolds, (**c**) conventional oven-annealed (CO 5 °C/min) nanofiber scaffolds and (**d**) YSZ bulk disk scaffolds. Representative images of actin cytoskeleton organisation of hMSCs on (**e**) microwave-annealed (MW 5 °C/min) scaffolds, (**f**) conventional oven-annealed (CO 5 °C/min) scaffolds and (**g**) YSZ bulk disk scaffolds (actin filaments labelled with phalloidin are shown in green, and nuclei labelled with DAPI are shown in blue). Images used from Gazquez et al. [97], reprinted with permission from ACS, copyright© 2016.

Additionally, Esfahani et al., conducted electrospinning of polyamide 6(PA6)/HA on the surface of the ZTA nanocomposites with improved bioactivity suitable for bone-repair applications [99]. Even though this study did not involve electrospun scaffolds, the embedding of 6(PA6)/HA is an innovative strategy that can be used in future research involving the fabrication of electrospun 6(PA6)/HA/zirconia scaffolds. Later, various researchers investigated different precursors and design strategies for fabricating flexible zirconia scaffolds via electrospinning [100–102]. Even though they successfully spun zirconia scaffolds and claimed that they are potentially suitable for bone-tissue engineering, there is a lack of in vitro and in vivo studies supporting these claims.

4. Hybrid Techniques

In contrast to the foregoing replica techniques, Feng et al., designed an acrylonitrile butadiene styrene (ABS) template with organised channel openings using a 3D Rapid Prototyper to construct zirconia scaffolds with different shapes (cubic and cylindrical) and porosities [103]. The authors coated the zirconia scaffolds with mesoporous bioglass (MBG) to enhance their bioactivity. For the cubic and cylindrical samples, the MBG coating increased the compressive strength from 34.22 to 55.01 MPa and from 44.35 to 123.32 MPa, respectively. The cell proliferation of BMSCs on coated and uncoated samples indicated better biocompatibility.

Sapkal et al., developed a biocomposite scaffold by infusing 10%, 20%, 30%, 40% and 50% zirconia with a β-TCP background via an indirect casting technique. The ABS was first printed using a conventional 3D printer, followed by the impregnation of a zirconia/β-TCP ceramic slurry and subsequent annealing to obtain a zirconia/β-TCP bioceramic scaffold. Osteosarcoma cell line (MG-63) results and compressive-testing results for a sample with a zirconia content of 30% indicated that it was suitable for bone-tissue engineering. Scaffold fabrication with proper porosities suitable for cortical and cancellous bone must be verified [104].

In the aforementioned studies, the direct fabrication of flexible zirconia-containing composite scaffolds via electrospinning was demonstrated. However, a pioneering methodology combining the traditional sponge replica and modern electrospinning methods for fabricating an artificial small bone was reported by Lee et al., in 2011 [105]. Zirconia-BCP/polymethylmethacrylate-PCL-HA (zirconia-BCP/PMMA-PCL-HA) was engineered by combining the sponge replica and electrospinning techniques. First, to mimic the cancellous bone, the authors fabricated a zirconia/BCP scaffold with an interconnected porous structure via the conventional sponge replica technique (zirconia and BCP were prepared separately for comparison). To mimic the Haversian canal zone of the bone, PMMA-PCL-HA nanofibers were fenced around a steel wire with a diameter of 0.3 mm. Finally, the osteon-like flexible fibers were collected and wrapped onto the zirconia/BCP scaffold and again electrospun with PMMA-PCL-HA fibers to produce the desired scaffold (Figure 14). The outstanding cell sustainability properties (MG-63 osteoblast-like cells) confirmed the cytocompatibility of the intentionally fabricated artificial bone (zirconia-BCP/PMMA-PCL-HA scaffold). The synthetic bone-like scaffold proposed by Kim et al., that aims to replicate finger and toe bones represents a significant pioneering achievement but requires further in vivo testing before clinical implementation.

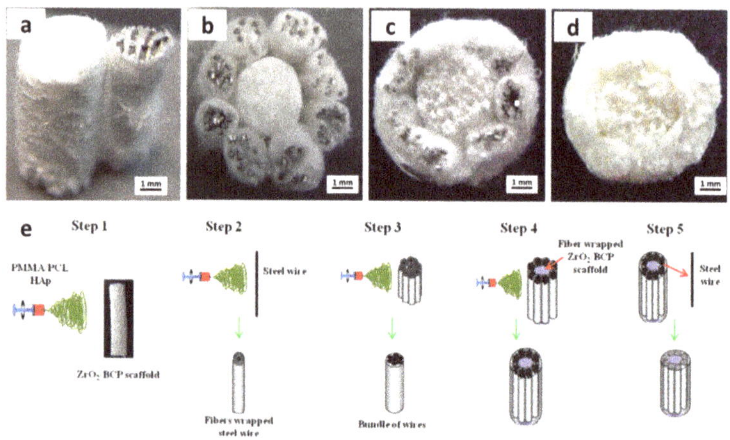

Figure 14. Photographs (**a**–**d**) and schematic (**e**) of the production of artificial bone. (**a**) A bundle of wires attached to the zirconia/BCPscaffold; (**b**) zirconia/BCP scaffold surrounded by bundles; (**c**) bundles surrounded by PCL/PMMA/HA; (**d**) after the steel wires were removed. Reprinted with permission from Lee et al. [105], copyright© 2011.

Recently, Sakthiabirami et al., fabricated porous zirconia 3D scaffolds via FDM [106]. The scaffolds were subjected to composite coating (Zn-HA/glass) on glass-impregnated zirconia to endorse hybrid roles with good mechanical and biological properties. Furthermore, the constructed scaffold was embedded with a biopolymer (alginate/gelatin) to obtain the ECM of the bone prototype (Figure 15a). The compression strength of the composite-coated scaffold and the biopolymer embedded on the composite-coated scaffold was approximately 20% higher than that of the pristine zirconia scaffold. Interestingly, the biopolymer-surrounded zirconia scaffolds were able to retain their shape, whereas the zirconia scaffolds without the biopolymer shattered into pieces (Figure 15b). This significantly affected the toughness of the biopolymer-embedded scaffold (Figure 15c). In vitro studies of hybrid scaffolds using dental pulp cells (DPCs) revealed enriched cell adhesion and cell differentiation. The authors suggested that their dynamic hybrid approach is suitable for load-bearing bone restorations; however, this should be confirmed via in vivo tests.

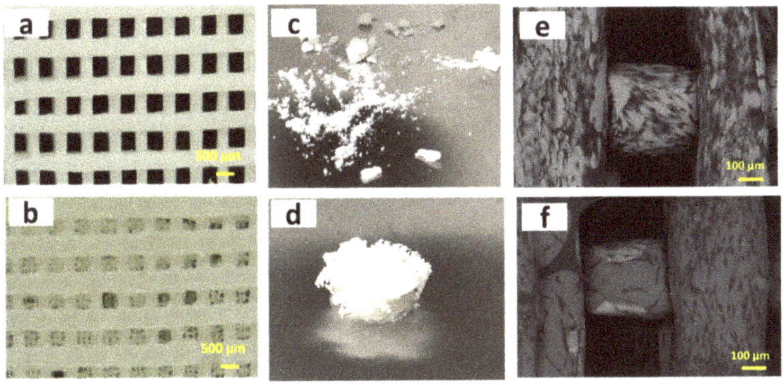

Figure 15. Microscopic images of (**a**) ZC and (**b**) ZCB. Digital photograph of (**c**) ZC and (**d**) ZCB after the compression test. Cell-attachment analysis using FE-SEM of the (**e**) Z and (**f**) ZC groups. Images used from Sakthiabirami et al. [106] reprinted with permission from Elsevier, copyright© 2021.

A summary of the zirconia-based scaffolds discussed in the previous sections that have been used for bone-regeneration applications over the past two decades is presented in Table 1. This table indicates not only the various types of scaffolds examined but also the composites/coatings used for the surface modifications. Since the bioactivity of zirconia-based scaffolds determines their clinical success, this review focused on the in vitro and in vivo outputs, along with the mechanical properties for load-bearing applications.

Very recently, Weng et.al summarized the properties and construction of ZrO_2 and its composite materials [107]. Moreover, we have observed substantial development that has been lately accomplished in this theme. However, a review article summarizing the development trend of zirconia-based scaffolds with respect to the evolution of 3D fabrication techniques is not documented yet so far. Thus, this review presents a complete summary of the designs and strategies of 3D fabrication techniques towards highly efficient 3D zirconia-based scaffolds centered on biological (in vivo/in vitro) and mechanical developments.

5. Prospects and Challenges

The non-degradability of zirconia-based scaffolds is a major bottleneck of zirconia-based in comparison with other calcium phosphate bioceramics-based bone-tissue engineering applications. Considerable research attention has been directed toward resolving this concern, and zirconia-based scaffolds have seen advancements over the decades owing to the repeated efforts of bone-tissue engineers. The successful long-term use of zirconia-based implants in dental applications has been validated via clinical trials with 90% survival rates. Even though zirconia-based ceramics are found to undergo fracture in the earlier stage, the absence of surplus parasitic reactions indicates that zirconia can cause no harm to the surrounding biological environments. Hence, zirconia-based scaffolds with 3D porous architectures can provide sufficient osteointegration; once, the bone-remodelling initiated earlier fracture will not be a major issue.

In this paper, we comprehensively reviewed the evolution of 3D zirconia-based scaffolds over the years, detailing different fabrication techniques, e.g., the conventional foam replica technique and modern 3D-printing methods. As indicated by the pie chart in Scheme 4a, the majority of the zirconia-based scaffolds were fabricated using the conventional sponge replication technique owing to its simplicity and rapidity. This cost-effective method produces the scaffolds with sufficient porosity and mechanical stability. However, the ability of the conventional sponge replication technique to construct scaffolds with the required porosity and a precise size and shape, which is critical for defining cell migration and tissue growth, is limited. Following the sponge replication technique, numerous conventional methods, including salt leaching, freeze-drying, foam replicas, solvent casting and particulate leaching, have been proposed for designing zirconia-based scaffolds.

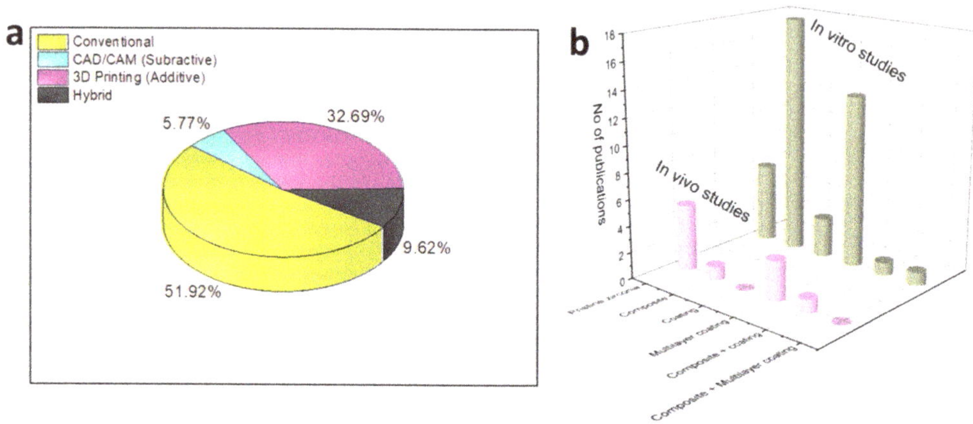

Scheme 4. (**a**) Assessment of the recent research articles on zirconia-based scaffolds based on the fabrication techniques, (**b**) comparison of the recent research articles on zirconia-based scaffolds based on in vivo and in vitro studies.

In the design aspects, pursuing the finest scaffold design for bone-repair applications has greatly increased; the recent evolution of 3D-printing techniques for designing 3D bioceramic scaffolds has encouraged bone-tissue engineers to construct 3D zirconia-based scaffolds with precise dimensions and interconnected porosity. Even though 3D printing is well established for calcium-based bioceramics aimed for bone-tissue engineering, few studies have been performed on 3D-printing techniques, including extrusion-based, DIW, DLP and SLS techniques, for fabricating zirconia-based scaffolds for bone-tissue engineering. This clearly indicates that 3D-printing techniques are still in a stage of infancy. Among the few studies reported thus far, a multi-extrusion technique-derived 3D zirconia-based scaffold exhibited considerable promise, with improved mechanical and biological properties. However, the surfactants, organic solvents and binders used for the 3D-printing process must be carefully selected, as they can be harmful to host cells and tissues. Given the limited number of reports on zirconia-based scaffolds fabricated via 3D-printing techniques, it may be challenging for future bone-tissue researchers to clearly understand the degradation, mechanical stability and bone-regeneration ability of 3D-printed zirconia scaffolds in load-bearing applications in human biological environments.

To address the poor interfacial compatibility between the zirconia scaffolds and incoming cells, three main strategies have been employed by bone-tissue engineers: surface reformation, interface-phase introduction and crystal-structure variation [108]. For instance, in the early 2000s, Lee et al., made a significant contribution to enhancing zirconia scaffolds for bone-regeneration applications by optimising diverse coating and fabrication techniques. The dual apatite FA/HA coating approach for nullifying the direct chemical transformation between the HA and zirconia scaffolds has been a pioneering outlook considered for future research. Additionally, bone-tissue researchers have included bioactive materials in zirconia scaffolds (resulting in composite scaffolds) for improving the bioactivity. Even though bioactive coating/composite formation increases the interfacial bonding strength, there are unresolved challenges to be addressed in future studies. For example, the inclusion of secondary phases in the zirconia scaffolds induces the chemical transformation of zirconia into undesired parasitic phases, which can reduce the strength [109]. Additionally, the mechanical and physical impacts of secondary bioactive coatings must be considered before practicing in bone defect locations.

Finally, it is essential to discuss the clinical possibilities of zirconia-based scaffolds. The direct comparison of the strength of the zirconia-based scaffolds to the strength of the natural bone (cortical bone) lacks among the reviewed articles, which is very essential to understand the ability of zirconia-based scaffolds to satisfy the critical bone defects requirements. In most of the collected literature, the in vitro studies focused on validating the metabolic activity of osteogenic cells (Scheme 4b). Surprisingly, there is a lack of in vivo proof of zirconia-based scaffolds in animal biological environments. Previous reports suggested that there is a large inconsistency between the in vitro performance and the in vivo performance [110]. In most of the reviewed literature on zirconia-based scaffolds, little attention was paid to the in vivo evolutions. In particular, zirconia-based scaffolds formulated via polymeric replication have been subjected to in vitro and in vivo evolutions. Indeed, polymeric replication-formulated zirconia-based scaffolds have been demonstrated to replicate hard-cortical and spongy-cancellous bone.

Few 3D scaffolding technologies have been commercially available, among which CAD/CAM is commonly used and the most dependable. After the successful fabrication of zirconia-based scaffolds for dental applications using the CAD/CAM technique. Apart from the in vitro and in vivo performance, 3D zirconia-based scaffolds fabricated using the CAD/CAM technique has been successfully employed for human mandibular reconstruction. Even though 3D-printed zirconia-based scaffolds were claimed to be the best possible alternatives to the conventional technologies, few in vivo studies support the clinical possibilities of 3D-printed zirconia-based scaffolds. Hence, it is essential to closely monitor the in-depth reproduction of natural mechanical and biological inducements and the rate of vascularisation of 3D-printed zirconia-based scaffolds in the in vivo environment. Multifaceted research should be targeted to systematically appreciate the perspective of 3D printing in bone-restoration applications.

Table 1. A summary of the zirconia based scaffolds for bone tissue engineering applications based on fabrication/design and its properties.

Year	Materials	Ceramic Content	Fabrication Techniques	Composite Materials	Infiltration or Intermediate Layer	Coating Materials	Porosity (%)	Pore Size(μm)	Mechanical Properties			Biological Properties		Ref.
									CS (MPa)	Modulus	Others	In Vitro	In Vivo (weeks)	
Conventional Technique														
2003	YSZ-FA-HA	-	Sponge replica	-	FA (~5 μm)	HA (~20 μm)	74–92	~600	1.6–35	-	Adhesion strength ~ 22 MPa	HOS cells	-	[42]
2004	YSZ-Ca P	-	Sponge replica	-	FA	Ca P (HA,FA,TCP,HA+FA,HA+TCP)	~90	500–700	-	-	-	SBF, MG-63 and HOS cells	-	[43]
2004	YSZ (5.2 wt% Y)-FA-HA	-	Sponge replica	-	FA	HA	90	~600	-	-	Adhesion strength ~ 33 MPa	Saline and MG-63 cells	-	[44]
2008	YSZ-FA-HA	-	Sponge replica	-	FA	HA	~84–87	~500–700 and ~150–200	~7-8	-	-	-	Rabbit calvarial defect (4,12)	[45]
2010	YSZ—MgO doped alumina	60 wt% (~20 vol%)	Pore former (Expanded polystyrene beads) and vacuum slip casting	YSZ (5, 10, 20 vol%)	-	58S33C or 58S bioglass (15–20 μm)	-	262–121	7.52–5.42	1.64–1.17 GPa	-	SBF	-	[63]
2011	YSZ-BCP	YSZ (10 vol%)	Sponge replica	-	BCP/YSZ (1:1)	BCP (top layer)	68.3	100–250	7.2	-	-	MG-63 cells	-	[50]
2012	YSZ-HA	-	Sponge replica	HA (20, 30, 40, 50 wt%)	-	Fibrin gel (with BMSCs encapsulation)	72–91	-	2.5–13.8	-	-	MCT3-E1 and BMSC cells	Rat calvarial defect (3,6)	[20]
2013	YSZ-bioglass-HA	-	Freeze drying	HA (18%)	45S5 bioglass	HA (~20 μm)	68.2	100–500	2.11	-	Flexural strength 343 MPa; Vickers hardness (substrate) 5.22 GPa	-	-	[59]
2013	YSZ-HA	-	Pore former (PMMA beads and starch powder)	HA (30 vol%)	1st layer-YSZ (20 and 10 vol%) and 2nd layer-HA+YSZ(5 vol.%) $+Al_2O_3$ (5 vol.%)	HA (pure)	-	1–10 and 20–50	-	-	Bending strength 320–390Mpa and Adhesion strength 24.5 MPa	L929 cells	-	[111]
2014	YSZ-Ca P	-	Sponge replica	β-TCP (50, 60, 70wt%)	-	-	65–84	-	4.95–6.25	48–63 MPa	-	h-ESC cells	-	[48]
2014	YSZ- β-TCP-HA	-	Sponge replica	-	-	β-TCP, HA and BCP	-	200–500	-	-	-	MC3T3-E1 cells	-	[51]
2015	YSZ-BCP-Collagen	-	Sponge replica	-	BCP/YSZ (1:1) (~10–15 μm)	BCP (top layer) (~10–15 μm) and collagen	-	150–500	6.8	-	-	MC3T3-E1 cells	Rabbit femoral defect (1 and 5 months)	[52]
2015	YSZ	-	Sponge replica	-	-	-	92.7–68	830–577	0.6–4.4	-	-	SBF and BMSC cells	-	[57]
2015	YSZ-Al_2O_3	-	Pore former (polyethylene beads)	Al_2O_3 (20wt%)	-	-	50–61	113–185	60–240	3–10 GPA	-	MC3T3-E1 cells	-	[112]
2015	YSZ and MgSZ	-	Pore former (polyethylene beads)	Y_2O_3 (8, 3wt%) and MgO (8, 3wt%)	-	-	50 and 57	167 and 144	210 and 120	10 and 6	-	MC3T3-E1 cells	-	[64]
2015	ZrO_2-CS-SF	-	Freeze drying	ZrO_2 (1)-CS (2)-SF(2.5)wt%	-	-	>75	50–150	-	-	-	PBS and HGF cells	-	[61]
2015	ZrO_2- POM-SF	-	Freeze drying	POM-ZrO_2-SF	-	-	-	100–200	-	-	-	SBF and HGF cells	-	[60]

Table 1. *Cont.*

Year	Materials	Ceramic Content	Fabrication Techniques	Composite Materials	Infiltration or Intermediate Layer	Coating Materials	Porosity (%)	Pore Size(μm)	Mechanical Properties			Biological Properties		Ref.
									CS (MPa)	Modulus	Others	In Vitro	In Vivo (weeks)	
2017	YSZ-HA-PLA	–	Solvent casting and salt-leaching	PLA-HA (10,15,20wt%)-YSZ (10, 15, 20 wt%)	–	–	–	–	0.8–6.9	0.7–5.1 MPa	Flexural strength 0.09–0.47 MPa; Flexural strain 17.5–8.2%	SBF	–	[66]
2018	YSZ	–	Sponge replica(crushed into granules after sintering)	–	–	–	–	300–400	–	–	–	–	Rabbit calvarial defect (2, 4, 8)	[27]
2018	YSZ-FA-HA	55	Sponge replica	–	FA (~45.7μm)	HA-PRP/HS	71.6	310 ± 150	2.98–18.8	–	–	MG-63cells	–	[46]
2018	ZrO_2-HA-PRP/HS	–	Sponge replica	–	FA	HA and Loaded with PRP gel/HS	–	–	–	–	–	–	Rabbit mandible defect (6, 8)	[47]
2018	ZrO_2	–	Solvent casting and Salt-leaching	–	–	–	79	100	–	–	–	–	Case study (repair of oroantral fistula)	[67]
2018	ZrO_2-CS-HA -CZ	–	Freeze drying	CS(2.5 wt%)-HA- ZrO_2 -CZ	–	–	85–94	10–100	0.3–1	CM-0.24–0.97 MPa; EM-3–4 GPa	–	PBS and OB-6 cells	–	[62]
2019	YSZ	50–70 wt%	Sponge replica	–	–	58S bioglass (~2μm)	94–85	700–322	–	–	–	MG-63cells	–	[55]
2020	YSZ	60wt%	Sponge replica	–	–	–	94–85	700–120	0.2–0.8	9–39 MPa	FEA, hardness (387 MPa)	–	–	[58]
2020	YSZ-58S	60 wt%	Sponge replica	–	58S bioglass (infiltration)	58S bioglass	93–89	–	0.33–0.44	–	–	SBF and MG-63 cells	–	[56]
2020	YSZ-HA-TCP	55 wt%	Sponge replica	YSZ (5, 10, 15, 20 and 25 wt%)	–	–	65.7–84.4	–	13.2–4.5	–	–	SBF	–	[49]
						Extrusion Based Printing								
2011	YSZ—HA(α-TCP)	48–43 vol%	Multipass extrusion	–	HA (α-TCP)—YSZ	HA	77	86	53	–	–	MG-63 cells	–	[79]
2012	YSZ—HA	45–40 vol%	Multipass extrusion	–	HA—YSZ	HA	–	–	7–20	–	–	MG-63 cells	–	[77]
2012	YSZ—BCP	46–41 vol%	Multipass extrusion	–	YSZ—BCP	PCL/BCP	92–78	–	8.27–12.7	–	–	MG-63 cells	–	[78]
2014	YSZ	70 wt%	Direct ink writing (DIW)	–	–	–	55 and 63	–	8 and 10	–	–	HCT116 cells	–	[23]
2017	ZrO_2—β-TCP	–	3D Bio-plotter	ZrO_2 (30wt%)	–	–	60–76.46	160–226	7–12.025	–	–	MG-63 cells	–	[79]
2018	YSZ-Al_2O_3(ZTA)	70wt% (35.5 vol%)	Robocasting	ZTA (YSZ-16 wt.%)	–	–	50	245	–	–	–	HOB cells	–	[87]
2018	ZrO_2-BCP	–	FDM	ZrO_2 (10 wt%)	–	–	–	350	0.5	–	–	MG-63 and hMSCs cells	–	[81]
2019	YSZ	48 vol%	Robocasting	–	–	–	–	200–500	–	–	–	–	–	[85]
2019	ZrO_2—β-Ca_2SiO_4	–	3D Bio-plotter	ZrO_2 (5, 10, 15 wt%)	–	–	~67	–	3.9–6.1	–	–	SBF and BMSC cells	Rat calvarial defect (8)	[80]
2020	ZrO_2—PCL	–	FDM	ZrO_2 (5, 10, 20 wt%)	–	–	46.2–47	459.2–462.7	5.5–7.9	43–67 MPa	–	MC3T3-E1 cells	–	[83]
2021	YSZ-FA-HA	39.5 vol%	Direct ink writing (DIW)	–	FA	HA (~20μm)	61.1–75.3	–	20.8–62.9	–	–	SBF	–	[86]

Table 1. *Cont.*

Year	Materials	Ceramic Content	Fabrication Techniques	Composite Materials	Infiltration or Intermediate Layer	Coating Materials	Porosity (%)	Pore Size(µm)	Mechanical Properties			Biological Properties		Ref.
									CS (MPa)	Modulus	Others	In Vitro	In Vivo (weeks)	
Polymerization Based Printing Technique														
2019	ZrO_2 -HA	60 wt%	DLP	ZrO_2 (1, 3, 6wt%)	–	–	–	–	–	–	Tensile strength (29.4%); Bending strength (23.9%)	BMSC cells	–	[26]
2020	YSZ-HA	–	DLP	HA (10, 20, 30 wt%)	–	–	54.6	–	52.25	–	CS after soaking in SBF (25 MPa)	SBF and MC3T3-E1 cells	–	[91]
CAD/CAM Technique														
2019	YSZ	–	CAD/CAM (5-axis milling)	–	–	–	45.06	–	52.25	–	FEA	SBF	–	[69]
2019	YSZ	–	CAD/CAM	–	–	–	–	–	–	–	–	–	Case study (mandibu-lar defects)	[68]
2021	YSZ	–	CAD/CAM	–	–	–	–	–	–	–	–	MC3T3-E1 cells	Rat femur (2,4,8)	[70]
Electrospinning														
2016	YSZ- PVP	–	Electrospinning	–	–	–	–	–	–	1.11 MPa	–	HMSCcells	–	[97]
2017	ZrO_2-PCL	6–30 wt.%	Electrospinning	–	–	–	–	–	–	–	–	3T3 cells	–	[98]
SLS Technique														
2014	ZrO_2-$CaSiO_3$	–	SLS	ZrO_2 (10, 20, 20, 40wt%)	–	–	70	1600	17.9–44.1	–	Fracture toughness 1.14–1.66 MPa.$m^{1/2}$	SBF and MG-63 cells	–	[93]
Hybrid Technique														
2011	YSZ-TCP	–	3D Rapid Prototyper (ABS template)followed by slurry impergation	–	–	Mesoporous bioglass	63–68	500–800	44.35–123.32	–	–	SBF and BMSC cells	–	[103]
2011	YSZ-BCP and PMMA-PCL-HA (Fiber)	10 vol%	Sponge replicaAnd Electrospinning	–	YSZ-BCP	BCP	67.68–69.65	–	4.83–4.97	–	–	MG-63 cells	–	[105]
2012	ZrO_2	50 vol.%	Free-form	–	–	–	40	350	–	–	–	–	Case study (maxilla)	[113]
2016	ZrO_2—β-TCP	–	3D Rapid Prototyper (ABS template) followed by impergation	ZrO_2 (10, 20, 30, 40, 50wt%)	–	–	68.5–82.5	–	3–15	184–396 MPa	–	PBS and MG-63 cells	–	[104]
2020	YSZ-bioglass-Zn-HA-Biopolymer	40 Vol%	FDM and Freeze drying	–	Glass (Infiltration)	Glass/Zn-HA(~1 µm) and Gelatin/alginate	~40%	300–450	68.2–89.8	1.7–2.6 GPa	Strain energy density 1.8-4.2MJ/m^3	DPCs cells	–	[106]

6. Conclusions

The primary goal of the scaffold is to get completely absorbed and replaced by body tissue. However, in large defects, scaffold or implanted artificial bone materials like HA or TCP are not absorbed completely and remained for a long time in vivo studies. And the healing period of each bone is different from several weeks to several months (ex. Femur 6 months). Until now researchers are not able to control the absorption rate of scaffold depending on new bone formation. Over the past 20 years, the zirconia-based scaffold has played a substantial role in critical-sized bone-defect applications, following the tremendous success of modern restorative dentistry. For the large sized loading area, a zirconia-based scaffold that has comparable strength to that of natural bone is needed to support bone formation during the long healing time. However, it remains challenging for orthopaedic experts and bone-tissue researchers to heal critical-sized bone deficiencies using zirconia-based scaffolds, owing to the immature fabrication techniques and unfavourable composite/coating materials for enhancing the bioactivity. Clinically feasible possibilities for promoting surface activation, along with appropriate 3D fabrication practices, are likely to be a focus area for future zirconia-based -scaffold design. The advancements in material engineering in association with biomechanical engineering could be critical in determining zirconia-based scaffolds as a successful biomaterial. Undoubtedly, with the expansion of advanced 3D-printing technology and unique coating strategies, the liability of zirconia-based scaffolds for critical-sized bone-replacement applications is not far away.

Author Contributions: Conceptualization and writing—original draft preparation, S.-W.P., K.S. and V.S.; writing—review and editing, S.-W.P., K.S. and V.S.; visualization, J.-H.K. and Y.P.Y.; supervision and funding acquisition, S.-W.P. All authors have read and agreed to the published version of the manuscript.

Funding: This work was supported by a National Research Foundation of Korea (NRF). Grant funded by the Korea government (MSIT) (No.2019R1A2C1089456).

Institutional Review Board Statement: Not applicable.

Informed Consent Statement: Not applicable.

Data Availability Statement: All the data were collected with copyrights permission.

Conflicts of Interest: The authors declare no conflict of interest.

Abbreviations

YSZ—3mol.% yttria stabilised zirconia, FA—Fluroappatite, HA—Hydroxyapatite, CaP—Calcium phosphate, TCP—Tricalcium phosphate, MgO—Magnesium oxide, BCP—Bicalcium phosphate, PMMA—Polymethamethylacralate, β-TCP-Beta—Tricalcium phosphate, Al2O3—Aluminium oxide/Alumina, MgSZ—Mgnesium stabilized zirconia, Y_2O_3—Yttrium oxide, ZrO_2—Zirconium dioxide/Zirconia, CS—Chitosan, SF—Silk fibrin, POM—Polyoxometalates, PLA—Polylatic acid, PRP—Plasma rich protein, HS—Heparin sulfate, CZ—Calcium zirconate, α-TCP-Apha—Tricalcium phosphate, ZTA—Alumina toughened zirconia, Ca_2SiO_4—Calcium silicate, PCL—Polycaprolactone, PVP—Polyvinylpyrrolidone, ABS—Acrylonitrile butadiene styrene, Zn-HA—Zinc doped hydroxyapatite, DIW—Direct ink writing, FDM—Fused deposition modelling, DLP—Digital light processing, CAD/CAM—Computer aided design/Computer aided milling, SLS—Selective laser sintering, CS—Compression strength, HOS—Human osteoscarcoma, SBF—Stimulated body fluid, MG63—Ostesarcoma cells, MCT3-E1—Murine preosteoblast cells, BMSC-Bone marrow—derived mesenchymal stem cells, L929—Murine fibroblast cells, PBS—Phosphate buffered saline, HGF—Human gingival fibroblast cells, OB6—Murine bone marrow-derived osteoblastic cells, HCT116—Human colon carcinoma cells, HOB—Human osteoblast cells, hMSC—Human mesenchymal stem cells, DPCs—Dental pulp cells.

References

1. Saridag, S.; Tak, O.; Alniacik, G. Basic properties and types of zirconia: An overview. *World J. Stomatol.* **2013**, *2*, 40. [CrossRef]
2. Abd El-Ghany, O.S.; Sherief, A.H. Zirconia based ceramics, some clinical and biological aspects: Review. *Future Dent. J.* **2016**, *2*, 55–64. [CrossRef]
3. Chen, Y.-W.; Moussi, J.; Drury, J.L.; Wataha, J.C. Zirconia in biomedical applications. *Expert Rev. Med. Devices* **2016**, *13*, 945–963. [CrossRef] [PubMed]
4. Christel, P.; Meunier, A.; Heller, M.; Torre, J.P.; Peille, C.N. Mechanical properties and short-term in vivo evaluation of yttrium-oxide-partially-stabilized zirconia. *J. Biomed. Mater. Res.* **1989**, *23*, 45–61. [CrossRef]
5. Heuer, A.H.; Lange, F.F.; Swain, M.V.; Evans, A.G. Transformation Toughening: An Overview. *J. Am. Ceram. Soc.* **1986**, *69*, i–iv. [CrossRef]
6. Zarone, F.; Russo, S.; Sorrentino, R. From porcelain-fused-to-metal to zirconia: Clinical and experimental considerations. *Dent. Mater.* **2011**, *27*, 83–96. [CrossRef]
7. Madfa, A.A.; Al-Sanabani, F.A.; Al-Qudami, N.H.; Al-Sanabani, J.S.; Amran, A.G. Use of Zirconia in Dentistry: An Overview. *Open Biomater. J.* **2014**, 1–9. [CrossRef]
8. Cranin, A.N.; Schnitman, P.A.; Rabkin, M.; Dennison, T.; Onesto, E.J. Alumina and zirconia coated vitallium oral endosteal implants in beagles. *J. Biomed. Mater. Res.* **1975**, *9*, 257–262. [CrossRef]
9. Jung, R.E.; Zembic, A.; Pjetursson, B.E.; Zwahlen, M.; Thoma, S.D. Systematic review of the survival rate and the incidence of biological, technical, and aesthetic complications of single crowns on implants reported in longitudinal studies with a mean follow-up of 5 years. *Clin. Oral Implant. Res.* **2012**, *23*, 2–21. [CrossRef]
10. Jung, R.E.; Pjetursson, B.E.; Glauser, R.; Zembic, A.; Zwahlen, M.; Lang, N.P. A systematic review of the 5-year survival and complication rates of implant-supported single crowns. *Clin. Oral Implant. Res.* **2008**, *19*, 119–130. [CrossRef]
11. Luthardt, R.G.; Sandkuhl, O.; Reitz, B. Zirconia-TZP and alumina—Advanced technologies for the manufacturing of single crowns. *Eur. J. Prosthodont. Restor. Dent.* **1999**, *7*, 113–119.
12. Tinschert, J.; Natt, G.; Mautsch, W.; Augthun, M.; Spiekermann, H. Fracture resistance of lithium disilicate-, alumina-, and zirconia-based three-unit fixed partial dentures: A laboratory study. *Int. J. Prosthodont.* **2001**, *14*, 231–238.
13. Meyenberg, K.H.; Lüthy, H.; Schärer, P. Zirconia posts: A new all-ceramic concept for nonvital abutment teeth. *J. Esthet. Dent.* **1995**, *7*, 73–80. [CrossRef]
14. Kakehashi, Y.; Lüthy, H.; Naef, R.; Wohlwend, A.; Schärer, P. A new all-ceramic post and core system: Clinical, technical, and in vitro results. *Int. J. Periodontics Restor. Dent.* **1998**, *18*, 586–593.
15. Christel, P.; Meunier, A.; Dorlot, J.M.; Crolet, J.M.; Witvoet, J.; Sedel, L.; Boutin, P. Biomechanical compatibility and design of ceramic implants for orthopedic surgery. *Ann. N. Y. Acad. Sci.* **1988**, *523*, 234–256. [CrossRef]
16. Soon, G.; Pingguan-Murphy, B.; Lai, K.W.; Akbar, S.A. Review of zirconia-based bioceramic: Surface modification and cellular response. *Ceram. Int.* **2016**, *42*, 12543–12555. [CrossRef]
17. Yin, L.; Nakanishi, Y.; Alao, A.-R.; Song, X.-F.; Abduo, J.; Zhang, Y. A Review of Engineered Zirconia Surfaces in Biomedical Applications. *Procedia CIRP* **2017**, *65*, 284–290. [CrossRef]
18. Cionca, N.; Hashim, D.; Mombelli, A. Zirconia dental implants: Where are we now, and where are we heading? *Periodontol. 2000* **2017**, *73*, 241–258. [CrossRef]
19. Al-Amleh, B.; Lyons, K.; Swain, M. Clinical trials in zirconia: A systematic review. *J. Oral Rehabil.* **2010**, *37*, 641–652. [CrossRef]
20. An, S.-H.; Matsumoto, T.; Miyajima, H.; Nakahira, A.; Kim, K.-H.; Imazato, S. Porous zirconia/hydroxyapatite scaffolds for bone reconstruction. *Dent. Mater.* **2012**, *28*, 1221–1231. [CrossRef]
21. Parfenov, V.A.; Mironov, V.A.; Koudan, E.V.; Nezhurina, E.K.; Karalkin, P.A.; Pereira, F.D.A.S.; Petrov, S.V.; Krokhmal, A.A.; Aydemir, T.; Vakhrushev, I.V.; et al. Fabrication of calcium phosphate 3D scaffolds for bone repair using magnetic levitational assembly. *Sci. Rep.* **2020**, *10*, 4013. [CrossRef]
22. Ho, C.M.B.; Ng, S.H.; Yoon, Y.-J. A review on 3D printed bioimplants. *Int. J. Precis. Eng. Manuf.* **2015**, *16*, 1035–1046. [CrossRef]
23. Li, Y.; Li, L.; Li, B. Direct write printing of three-dimensional ZrO_2 biological scaffolds. *Mater. Des.* **2015**, *72*, 16–20. [CrossRef]
24. Wei, G.; Ma, P.X. Structure and properties of nano-hydroxyapatite/polymer composite scaffolds for bone tissue engineering. *Biomaterials* **2004**, *25*, 4749–4757. [CrossRef]
25. Kondo, N.; Ogose, A.; Tokunaga, K.; Ito, T.; Arai, K.; Kudo, N.; Inoue, H.; Irie, H.; Endo, N. Bone formation and resorption of highly purified beta-tricalcium phosphate in the rat femoral condyle. *Biomaterials* **2005**, *26*, 5600–5608. [CrossRef]
26. Zhang, J.; Huang, D.; Liu, S.; Dong, X.; Li, Y.; Zhang, H.; Yang, Z.; Su, Q.; Huang, W.; Zheng, W.; et al. Zirconia toughened hydroxyapatite biocomposite formed by a DLP 3D printing process for potential bone tissue engineering. *Mater. Sci. Eng. C* **2019**, *105*, 110054. [CrossRef]
27. Kim, I.-J.; Shin, S.-Y. Comparative study of new bone formation capability of zirconia bone graft material in rabbit calvarial. *J. Adv. Prosthodont.* **2018**, *10*, 167–176. [CrossRef]
28. Shang, L.; Yu, Y.; Liu, Y.; Chen, Z.; Kong, T.; Zhao, Y. Spinning and Applications of Bioinspired Fiber Systems. *ACS Nano* **2019**, *13*, 2749–2772. [CrossRef]
29. Du, X.; Fu, S.; Zhu, Y. 3D printing of ceramic-based scaffolds for bone tissue engineering: An overview. *J. Mater. Chem. B* **2018**, *6*, 4397–4412. [CrossRef]
30. Service, R.F. Tissue engineers build new bone. *Science* **2000**, *289*, 1498–1500. [CrossRef]

31. Jakus, A.E.; Rutz, A.L.; Jordan, S.W.; Kannan, A.; Mitchell, S.M.; Yun, C.; Koube, K.D.; Yoo, S.C.; Whiteley, H.E.; Richter, C.-P.; et al. Hyperelastic "bone": A highly versatile, growth factor–free, osteoregenerative, scalable, and surgically friendly biomaterial. *Sci. Transl. Med.* **2016**, *8*, 358ra127. [CrossRef] [PubMed]
32. Turnbull, G.; Clarke, J.; Picard, F.; Riches, P.; Jia, L.; Han, F.; Li, B.; Shu, W. 3D bioactive composite scaffolds for bone tissue engineering. *Bioact. Mater.* **2018**, *3*, 278–314. [CrossRef] [PubMed]
33. Trombetta, R.; Inzana, J.A.; Schwarz, E.M.; Kates, S.L.; Awad, H.A. 3D Printing of Calcium Phosphate Ceramics for Bone Tissue Engineering and Drug Delivery. *Ann. Biomed. Eng.* **2017**, *45*, 23–44. [CrossRef] [PubMed]
34. Zhang, K.; He, R.; Xie, C.; Wang, G.; Ding, G.; Wang, M.; Song, W.; Fang, D. Photosensitive ZrO_2 suspensions for stereolithography. *Ceram. Int.* **2019**, *45*, 12189–12195. [CrossRef]
35. Muhamad Nor, M.A.A.; Hong, L.C.; Arifin Ahmad, Z.; Md Akil, H. Preparation and characterization of ceramic foam produced via polymeric foam replication method. *J. Mater. Process. Technol.* **2008**, *207*, 235–239. [CrossRef]
36. Jayakumar, R.; Ramachandran, R.; Sudheesh Kumar, P.T.; Divyarani, V.V.; Srinivasan, S.; Chennazhi, K.P.; Tamura, H.; Nair, S.V. Fabrication of chitin–chitosan/nano ZrO_2 composite scaffolds for tissue engineering applications. *Int. J. Biol. Macromol.* **2011**, *49*, 274–280. [CrossRef]
37. Sun, F.; Zhou, H.; Lee, J. Various preparation methods of highly porous hydroxyapatite/polymer nanoscale biocomposites for bone regeneration. *Acta Biomater.* **2011**, *7*, 3813–3828. [CrossRef]
38. Sulaiman, S.B.; Keong, T.K.; Cheng, C.H.; Saim, A.B.; Idrus, R.B.H. Tricalcium phosphate/hydroxyapatite (TCP-HA) bone scaffold as potential candidate for the formation of tissue engineered bone. *Indian J. Med. Res.* **2013**, *137*, 1093–1101.
39. Deschamps, I.S.; Magrin, G.L.; Magini, R.S.; Fredel, M.C.; Benfatti, C.A.M.; Souza, J.C.M. On the synthesis and characterization of β-tricalcium phosphate scaffolds coated with collagen or poly (D, L-lactic acid) for alveolar bone augmentation. *Eur. J. Dent.* **2017**, *11*, 496–502. [CrossRef]
40. Kim, D.-H.; Kim, K.-L.; Chun, H.-H.; Kim, T.-W.; Park, H.-C.; Yoon, S.-Y. In vitro biodegradable and mechanical performance of biphasic calcium phosphate porous scaffolds with unidirectional macro-pore structure. *Ceram. Int.* **2014**, *40*, 8293–8300. [CrossRef]
41. Mostafa, A.A.; Oudadesse, H.; Mohamed, M.B.; Foad, E.S.; Le Gal, Y.; Cathelineau, G. Convenient approach of nanohydroxyapatite polymeric matrix composites. *Chem. Eng. J.* **2009**, *153*, 187–192. [CrossRef]
42. Kim, H.-W.; Lee, S.-Y.; Bae, C.-J.; Noh, Y.-J.; Kim, H.-E.; Kim, H.-M.; Ko, J.S. Porous ZrO_2 bone scaffold coated with hydroxyapatite with fluorapatite intermediate layer. *Biomaterials* **2003**, *24*, 3277–3284. [CrossRef]
43. Kim, H.-W.; Kim, H.-E.; Salih, V.; Knowles, J.C. Dissolution control and cellular responses of calcium phosphate coatings on zirconia porous scaffold. *J. Biomed. Mater. Res. A* **2004**, *68*, 522–530. [CrossRef]
44. Kim, H.-W.; Kim, H.-E.; Knowles, J.C. Hard-tissue-engineered zirconia porous scaffolds with hydroxyapatite sol-gel and slurry coatings. *J. Biomed. Mater. Res. B. Appl. Biomater.* **2004**, *70*, 270–277. [CrossRef]
45. Kim, H.-W.; Shin, S.-Y.; Kim, H.-E.; Lee, Y.-M.; Chung, C.-P.; Lee, H.-H.; Rhyu, I.-C. Bone formation on the apatite-coated zirconia porous scaffolds within a rabbit calvarial defect. *J. Biomater. Appl.* **2008**, *22*, 485–504. [CrossRef]
46. Latifi, M.; Talaei-Khozani, T.; Mehraban-Jahromi, H.; Sani, M.; Sadeghi-Atabadi, M.; Fazel-Anvari, A.; Kabir-Salmani, M. Fabrication of platelet-rich plasma heparin sulfate/hydroxyapatite/zirconia scaffold. *Bioinspired Biomim. Nanobiomater.* **2018**, *7*, 122–130. [CrossRef]
47. Shahsavari-Pour, S.; Aliabadi, E.; Latifi, M.; Zareifard, N.; Namavar, M.R.; Talaei-Khozani, T. Evaluation of the Possible Synergic Regenerative Effects of Platelet-Rich Plasma and Hydroxyapatite/Zirconia in the Rabbit Mandible Defect Model. *Iran. J. Med. Sci.* **2018**, *43*, 633–644.
48. Alizadeh, A.; Moztarzadeh, F.; Ostad, S.N.; Azami, M.; Geramizadeh, B.; Hatam, G.; Bizari, D.; Tavangar, S.M.; Vasei, M.; Ai, J. Synthesis of calcium phosphate-zirconia scaffold and human endometrial adult stem cells for bone tissue engineering. *Artif. Cells Nanomed. Biotechnol.* **2016**, *44*, 66–73. [CrossRef]
49. Mohammad, M.; Taha, A.; Shahatha, S. Preparation of hydroxyapatite/zirconia porous composites via polymeric sponge method and study the physical and bioactivity properties. In Proceedings of the IOP Conference Series: Materials Science and Engineering, Istanbul, Turkey, 20–25 September 2019; IOP Publishing: Bristol, UK, 2020; Volume 757, p. 012046. [CrossRef]
50. Kim, M.; Franco, R.A.; Lee, B.-T. Synthesis of functional gradient BCP/ZrO_2 bone substitutes using ZrO_2 and BCP nanopowders. *J. Eur. Ceram. Soc.* **2011**, *31*, 1541–1548. [CrossRef]
51. Song, Y.-G.; Cho, I.-H. Characteristics and osteogenic effect of zirconia porous scaffold coated with β-TCP/HA. *J. Adv. Prosthodont.* **2014**, *6*, 285–294. [CrossRef]
52. Linh, N.T.B.; Jang, D.-W.; Lee, B.-T. Collagen immobilization of multi-layered BCP-ZrO_2 bone substitutes to enhance bone formation. *Appl. Surf. Sci.* **2015**, *345*, 238–248. [CrossRef]
53. Jones, A.C.; Arns, C.H.; Sheppard, A.P.; Hutmacher, D.W.; Milthorpe, B.K.; Knackstedt, M.A. Assessment of bone ingrowth into porous biomaterials using MICRO-CT. *Biomaterials* **2007**, *28*, 2491–2504. [CrossRef]
54. Mačković, M.; Hoppe, A.; Detsch, R.; Mohn, D.; Stark, W.J.; Spiecker, E.; Boccaccini, A.R. Bioactive glass (type 45S5) nanoparticles: In vitro reactivity on nanoscale and biocompatibility. *J. Nanopart. Res.* **2012**, *14*, 966. [CrossRef]
55. Mesquita-Guimarães, J.; Ramos, L.; Detsch, R.; Henriques, B.; Fredel, M.C.; Silva, F.S.; Boccaccini, A.R. Evaluation of in vitro properties of 3D micro-macro porous zirconia scaffolds coated with 58S bioactive glass using MG-63 osteoblast-like cells. *J. Eur. Ceram. Soc.* **2019**, *39*, 2545–2558. [CrossRef]

56. Gouveia, P.F.; Mesquita-Guimarães, J.; Galárraga-Vinueza, M.E.; Souza, J.C.M.; Silva, F.S.; Fredel, M.C.; Boccaccini, A.R.; Detsch, R.; Henriques, B. In-vitro mechanical and biological evaluation of novel zirconia reinforced bioglass scaffolds for bone repair. *J. Mech. Behav. Biomed. Mater.* **2021**, *114*, 104164. [CrossRef]
57. Zhu, Y.; Zhu, R.; Ma, J.; Weng, Z.; Wang, Y.; Shi, X.; Li, Y.; Yan, X.; Dong, Z.; Xu, J.; et al. In vitro cell proliferation evaluation of porous nano-zirconia scaffolds with different porosity for bone tissue engineering. *Biomed. Mater.* **2015**, *10*, 55009. [CrossRef]
58. Askari, E.; Cengiz, I.F.; Alves, J.L.; Henriques, B.; Flores, P.; Fredel, M.C.; Reis, R.L.; Oliveira, J.M.; Silva, F.S.; Mesquita-Guimarães, J. Micro-CT based finite element modelling and experimental characterization of the compressive mechanical properties of 3-D zirconia scaffolds for bone tissue engineering. *J. Mech. Behav. Biomed. Mater.* **2020**, *102*, 103516. [CrossRef]
59. Jiang, B.; Hu, X.; Huang, Z. Porous bio-ceramic coating on zirconia formed through freeze-drying. *Mater. Lett.* **2013**, *109*, 66–69. [CrossRef]
60. Teimouri, A.; Ghorbanian, L.; Salavati, H.; Chermahini, A.N. Fabrication and characterization of POM/ZrO_2/silk fibroin composite scaffolds. *Mater. Lett.* **2015**, *157*, 85–88. [CrossRef]
61. Teimouri, A.; Ebrahimi, R.; Emadi, R.; Beni, B.H.; Chermahini, A.N. Nano-composite of silk fibroin–chitosan/Nano ZrO_2 for tissue engineering applications: Fabrication and morphology. *Int. J. Biol. Macromol.* **2015**, *76*, 292–302. [CrossRef] [PubMed]
62. Gaihre, B.; Jayasuriya, A.C. Comparative investigation of porous nano-hydroxyapaptite/chitosan, nano-zirconia/chitosan and novel nano-calcium zirconate/chitosan composite scaffolds for their potential applications in bone regeneration. *Mater. Sci. Eng. C* **2018**, *91*, 330–339. [CrossRef] [PubMed]
63. Liu, J.; Dong, Z.; Miao, X. Porous Alumina/Zirconia Composite Scaffold with Bioactive Glass 58S33C Coating. *J. Biomim. Biomater. Tissue Eng.* **2010**, *6*, 87–104. [CrossRef]
64. Hadjicharalambous, C.; Mygdali, E.; Prymak, O.; Buyakov, A.; Kulkov, S.; Chatzinikolaidou, M. Proliferation and osteogenic response of MC3T3-E1 pre-osteoblastic cells on porous zirconia ceramics stabilized with magnesia or yttria. *J. Biomed. Mater. Res. A* **2015**, *103*, 3612–3624. [CrossRef]
65. Perić Kačarević, Ž.; Rider, P.; Alkildani, S.; Retnasingh, S.; Pejakić, M.; Schnettler, R.; Gosau, M.; Smeets, R.; Jung, O.; Barbeck, M. An introduction to bone tissue engineering. *Int. J. Artif. Organs* **2020**, *43*, 69–86. [CrossRef]
66. Ziaee, F.; Zebarjad, S.M.; Javadpour, S. Compressive and flexural properties of novel polylactic acid/hydroxyapatite/yttria-stabilized zirconia hybrid nanocomposite scaffold. *Int. J. Polym. Mater. Polym. Biomater.* **2018**, *67*, 229–238. [CrossRef]
67. Mokhtar, K.E.; Elmahallawy, A.S.; Elashwah, A.A. Evaluation of the Use of Zirconia Porous Scaffold in the Repair of Oroantral Fistula. *Alex. Dent. J.* **2018**, *43*, 87–93. [CrossRef]
68. Aftan, K.T. 3D Reconstruction of Mandibular Defects using CAD/CAM Designed Zirconia Prosthesis. *Al Anbar Med. J.* **2019**, *15*, 24–29. [CrossRef]
69. Marques, A.; Miranda, G.; Faria, D.; Pinto, P.; Silva, F.; Carvalho, Ó. Novel design of low modulus high strength zirconia scaffolds for biomedical applications. *J. Mech. Behav. Biomed. Mater.* **2019**, *97*, 375–384. [CrossRef]
70. Petre, A.; Balta, C.; Herman, H.; Gharbia, S.; Codreanu, A.; Onita-Mladin, B.; Anghel-Zurbau, N.; Hermenean, A.-G.; Ignat, S.-R.; Dinescu, S.; et al. A novel experimental approach to evaluate guided bone regeneration (GBR) in the rat femur using a 3D-printed CAD/CAM zirconia space-maintaining barrier. *J. Adv. Res.* **2021**, *28*, 221–229. [CrossRef]
71. Kumar Gain, A.; Lee, B.-T. Fabrication of HAp-Coated Micro-Channelled t-ZrO_2 Bodies by the Multi-Pass Extrusion Process. *J. Am. Ceram. Soc.* **2006**, *89*, 2051–2056. [CrossRef]
72. Lee, B.-T.; Kim, K.-H.; Han, J.-K. Microstructures and material properties of fibrous Al_2O_3–(m-ZrO_2)/t-ZrO_2 composites fabricated by a fibrous monolithic process. *J. Mater. Res.* **2004**, *19*, 3234–3241. [CrossRef]
73. Lee, B.-T.; Kang, I.C.; Gain, A.K.; Kim, K.-H.; Song, H.-Y. Fabrication of pore-gradient Al_2O_3–ZrO_2 sintered bodies by fibrous monolithic process. *J. Eur. Ceram. Soc.* **2006**, *26*, 3525–3530. [CrossRef]
74. Gain, A.K.; Lee, B.-T. Microstructure control of continuously porous t-ZrO_2 bodies fabricated by multi-pass extrusion process. *Mater. Sci. Eng. A* **2006**, *419*, 269–275. [CrossRef]
75. Lee, B.-T.; Kang, I.-C.; Cho, S.-H.; Song, H.-Y. Fabrication of a Continuously Oriented Porous Al_2O_3 Body and Its In Vitro Study. *J. Am. Ceram. Soc.* **2005**, *88*, 2262–2266. [CrossRef]
76. Jang, D.-W.; Kim, Y.-H.; Lee, B.-T. Microstructure control of TCP/TCP-(t-ZrO_2)/t-ZrO_2 composites for artificial cortical bone. *Mater. Sci. Eng. C* **2011**, *31*, 1660–1666. [CrossRef]
77. Jang, D.-W.; Nguyen, T.-H.; Sarkar, S.K.; Lee, B.-T. Microwave sintering and in vitro study of defect-free stable porous multilayered HAp–ZrO_2 artificial bone scaffold. *Sci. Technol. Adv. Mater.* **2012**, *13*, 35009. [CrossRef]
78. Mondal, D.; So-Ra, S.; Sarkar, S.K.; Min, Y.K.; Yang, H.M.; Lee, B.T. Fabrication of multilayer ZrO_2–biphasic calcium phosphate–poly-caprolactone unidirectional channeled scaffold for bone tissue formation. *J. Biomater. Appl.* **2012**, *28*, 462–472. [CrossRef]
79. Sapkal, P.S.; Kuthe, A.M.; Mathankar, S.; Deshmukh, A.A. 3D Bio-Plotted Tricalcium Phosphate/Zirconia Composite Scaffolds to Heal Large Size Bone Defects. *Mol. Cell. Biomech.* **2017**, *14*, 125–136. [CrossRef]
80. Fu, S.Y.; Yu, B.; Ding, H.F.; Shi, G.D.; Zhu, Y.F. Zirconia Incorporation in 3D Printed β-Ca_2SiO_4 Scaffolds on Their Physicochemical and Biological Property. *J. Inorg. Mater.* **2019**, *34*, 444–454. [CrossRef]
81. Sa, M.-W.; Nguyen, B.-N.B.; Moriarty, R.A.; Kamalitdinov, T.; Fisher, J.P.; Kim, J.Y. Fabrication and evaluation of 3D printed BCP scaffolds reinforced with ZrO_2 for bone tissue applications. *Biotechnol. Bioeng.* **2018**, *115*, 989–999. [CrossRef]

82. Catauro, M.; Bollino, F.; Papale, F.; Mozetic, P.; Rainer, A.; Trombetta, M. Biological response of human mesenchymal stromal cells to titanium grade 4 implants coated with PCL/ZrO_2 hybrid materials synthesized by sol–gel route: In vitro evaluation. *Mater. Sci. Eng. C* **2014**, *45*, 395–401. [CrossRef]
83. Wang, Q.; Ma, Z.; Wang, Y.; Zhong, L.; Xie, W. Fabrication and characterization of 3D printed biocomposite scaffolds based on PCL and zirconia nanoparticles. *Bio Des. Manuf.* **2021**, *4*, 60–71. [CrossRef]
84. Peng, E.; Wei, X.; Garbe, U.; Yu, D.; Edouard, B.; Liu, A.; Ding, J. Robocasting of dense yttria-stabilized zirconia structures. *J. Mater. Sci.* **2018**, *53*, 247–273. [CrossRef]
85. Brazete, D.; Neto, A.S.; Ferreira, J.M.F. Optimization of zirconia inks to fabricate 3D porous scaffolds by robocasting. *Lek. Tech. Clin. Technol.* **2019**, *49*, 5–10. [CrossRef]
86. Kocyło, E.; Franchin, G.; Colombo, P.; Chmielarz, A.; Potoczek, M. Hydroxyapatite-coated ZrO_2 scaffolds with a fluorapatite intermediate layer produced by direct ink writing. *J. Eur. Ceram. Soc.* **2021**, *41*, 920–928. [CrossRef]
87. Stanciuc, A.-M.; Sprecher, C.M.; Adrien, J.; Roiban, L.I.; Alini, M.; Gremillard, L.; Peroglio, M. Robocast zirconia-toughened alumina scaffolds: Processing, structural characterisation and interaction with human primary osteoblasts. *J. Eur. Ceram. Soc.* **2018**, *38*, 845–853. [CrossRef]
88. Gabriel, S.; Hull, C.W. Apparatus for Production of Three-Dimensional Objects by Stereolithography. U.S. Patent 4575330A, 8 August 1984.
89. Chen, Z.; Li, Z.; Li, J.; Liu, C.; Lao, C.; Fu, Y.; Liu, C.; Li, Y.; Wang, P.; He, Y. 3D printing of ceramics: A review. *J. Eur. Ceram. Soc.* **2019**, *39*, 661–687. [CrossRef]
90. Chen, F.; Zhu, H.; Wu, J.-M.; Chen, S.; Cheng, L.-J.; Shi, Y.-S.; Mo, Y.-C.; Li, C.-H.; Xiao, J. Preparation and biological evaluation of ZrO_2 all-ceramic teeth by DLP technology. *Ceram. Int.* **2020**, *46*, 11268–11274. [CrossRef]
91. Cao, Y.; Shi, T.; Jiao, C.; Liang, H.; Chen, R.; Tian, Z.; Zou, A.; Yang, Y.; Wei, Z.; Wang, C.; et al. Fabrication and properties of zirconia/hydroxyapatite composite scaffold based on digital light processing. *Ceram. Int.* **2020**, *46*, 2300–2308. [CrossRef]
92. Savalani, M.M.; Hao, L.; Zhang, Y.; Tanner, K.E.; Harris, R.A. Fabrication of porous bioactive structures using the selective laser sintering technique. *Proc. Inst. Mech. Eng. Part. H J. Eng. Med.* **2007**, *221*, 873–886. [CrossRef]
93. Shuai, C.; Feng, P.; Yang, B.; Cao, Y.; Min, A.; Peng, S. Effect of Nano-Zirconia on the Mechanical and Biological Properties of Calcium Silicate Scaffolds. *Int. J. Appl. Ceram. Technol.* **2015**, *12*, 1148–1156. [CrossRef]
94. Yu, Z.; Liu, B.; Zhou, H.; Feng, C.; Wang, X.; Yuan, K.; Gan, X.; Zhu, L.; Zhang, G.; Xu, D. Mesoporous ZrO_2 fibers with enhanced surface area and the application as recyclable absorbent. *Appl. Surf. Sci.* **2017**, *399*, 288–297. [CrossRef]
95. Nemati, S.; Kim, S.-J.; Shin, Y.M.; Shin, H. Current progress in application of polymeric nanofibers to tissue engineering. *Nano Converg.* **2019**, *6*, 36. [CrossRef]
96. Zafar, M.S.; Najeeb, S.; Khurshid, Z.; Vazirzadeh, M.; Zohaib, S.; Najeeb, B.; Sefat, F. Potential of Electrospun Nanofibers for Biomedical and Dental Applications. *Materials* **2016**, *9*, 73. [CrossRef]
97. Cadafalch Gazquez, G.; Chen, H.; Veldhuis, S.A.; Solmaz, A.; Mota, C.; Boukamp, B.A.; van Blitterswijk, C.A.; ten Elshof, J.E.; Moroni, L. Flexible Yttrium-Stabilized Zirconia Nanofibers Offer Bioactive Cues for Osteogenic Differentiation of Human Mesenchymal Stromal Cells. *ACS Nano* **2016**, *10*, 5789–5799. [CrossRef]
98. Thakare, V.G.; Joshi, P.A.; Godse, R.R.; Bhatkar, V.B.; Wadegaokar, P.A.; Omanwar, S.K. Fabrication of polycaprolactone/zirconia nanofiber scaffolds using electrospinning technique. *J. Polym. Res.* **2017**, *24*, 232. [CrossRef]
99. Esfahani, H.; Darvishghanbar, M.; Farshid, B. Enhanced bone regeneration of zirconia-toughened alumina nanocomposites using PA6/HA nanofiber coating via electrospinning. *J. Mater. Res.* **2018**, *33*, 4287–4295. [CrossRef]
100. Ghelich, R.; Mehdinavaz Aghdam, R.; Sadat Torknik, F.; Reza Jahannama, M. Synthesis and characterization of biocompatible zirconia nanofibers based on electrospun PVP/$Zr(OPr)_4$. *Mater. Today Proc.* **2018**, *5*, 15733–15738. [CrossRef]
101. Castkova, K.; Maca, K.; Sekaninova, J.; Nemcovsky, J.; Cihlar, J. Electrospinning and thermal treatment of yttria doped zirconia fibres. *Ceram. Int.* **2017**, *43*, 7581–7587. [CrossRef]
102. Rodaev, V.V.; Razlivalova, S.S.; Zhigachev, A.O.; Vasyukov, V.M.; Golovin, Y.I. Preparation of Zirconia Nanofibers by Electrospinning and Calcination with Zirconium Acetylacetonate as Precursor. *Polymers* **2019**, *11*, 1067. [CrossRef] [PubMed]
103. Lin, F.; Yan, C.; Zheng, W.; Fan, W.; Adam, C.; Oloyede, A. Preparation of Mesoporous Bioglass Coated Zirconia Scaffold for Bone Tissue Engineering. *Adv. Mater. Res.* **2012**, *365*, 209–215. [CrossRef]
104. Sapkal, P.S.; Kuthe, A.M.; Kashyap, R.S.; Nayak, A.R.; Kuthe, S.A.; Kawle, A.P. Indirect casting of patient-specific tricalcium phosphate zirconia scaffolds for bone tissue regeneration using rapid prototyping methodology. *J. Porous Mater.* **2017**, *24*, 1013–1023. [CrossRef]
105. Kim, Y.-H.; Lee, B.-T. Novel approach to the fabrication of an artificial small bone using a combination of sponge replica and electrospinning methods. *Sci. Technol. Adv. Mater.* **2011**, *12*, 35002. [CrossRef]
106. Sakthiabirami, K.; Kang, J.-H.; Jang, J.-G.; Soundharrajan, V.; Lim, H.-P.; Yun, K.-D.; Park, C.; Lee, B.-N.; Yang, Y.P.; Park, S.-W. Hybrid porous zirconia scaffolds fabricated using additive manufacturing for bone tissue engineering applications. *Mater. Sci. Eng. C* **2021**, *123*, 111950. [CrossRef]
107. Weng, W.; Wu, W.; Hou, M.; Liu, T.; Wang, T.; Yang, H. Review of zirconia-based biomimetic scaffolds for bone tissue engineering. *J. Mater. Sci.* **2021**, *56*, 8309–8333. [CrossRef]
108. Feng, P.; He, J.; Peng, S.; Gao, C.; Zhao, Z.; Xiong, S.; Shuai, C. Characterizations and interfacial reinforcement mechanisms of multicomponent biopolymer based scaffold. *Mater. Sci. Eng. C* **2019**, *100*, 809–825. [CrossRef]

109. Sakthiabirami, K.; Vu, V.T.; Kim, J.W.; Kang, J.H.; Jang, K.J.; Oh, G.J.; Fisher, J.G.; Yun, K.D.; Lim, H.P.; Park, S.W. Tailoring interfacial interaction through glass fusion in glass/zinc-hydroxyapatite composite coatings on glass-infiltrated zirconia. *Ceram. Int.* **2018**, *44*, 16181–16190. [CrossRef]
110. Asadian, M.; Chan, K.V.; Norouzi, M.; Grande, S.; Cools, P.; Morent, R.; De Geyter, N. Fabrication and Plasma Modification of Nanofibrous Tissue Engineering Scaffolds. *Nanomaterials* **2020**, *10*, 119. [CrossRef]
111. Yang, J.-Z.; Sultana, R.; Hu, X.-Z.; Ichim, P. Novel Layered Hydroxyapatite/Tri-Calcium Phosphate–Zirconia Scaffold Composite with High Bending Strength for Load-Bearing Bone Implant Application. *Int. J. Appl. Ceram. Technol.* **2014**, *11*, 22–30. [CrossRef]
112. Hadjicharalambous, C.; Buyakov, A.; Buyakova, S.; Kulkov, S.; Chatzinikolaidou, M. Porous alumina, zirconia and alumina/zirconia for bone repair: Fabrication, mechanical and in vitro biological response. *Biomed. Mater.* **2015**, *23*, 025012. [CrossRef]
113. Malmström, J.; Slotte, C.; Adolfsson, E.; Norderyd, O.; Thomsen, P. Bone response to free form-fabricated hydroxyapatite and zirconia scaffolds: A histological study in the human maxilla. *Clin. Oral Implants Res.* **2009**, *20*, 379–385. [CrossRef] [PubMed]

materials

MDPI

Article

The Impact of Particle Size and Surface Treatment of Zirconia Suspension for Photocuring Additive Manufacturing

Mee-Jin Jun [1,†], Jin-Ho Kang [2,†], Kumaresan Sakthiabirami [2,3], Seyed Aliakbar Hosseini Toopghara [4], Ye-Seul Kim [2], Kwi-Dug Yun [2] and Sang-Won Park [2,3,4,*]

1 Department of Dental Hygiene, Gwangju Health University, Gwangju 62287, Republic of Korea
2 Department of Prosthodontics, School of Dentistry, Chonnam National University, Gwangju 61186, Republic of Korea
3 Biomedical Evaluation and Research Centre, School of Dentistry, Chonnam National University, Gwangju 61186, Republic of Korea
4 Department of Medical Engineering Joint Research, Chonnam National University, Gwangju 61186, Republic of Korea
* Correspondence: psw320@chonnam.ac.kr; Tel.: +82-62-530-5842
† These authors contributed equally to this.

Abstract: To prepare a photocurable ceramic suspension for use in commercialized additive manufacturing equipment, the effects of the rheological properties of zirconia particles added to a binder, and the presence or absence of a silane coupling agent on the particles was evaluated. To this end, three experimental groups (ZSs, ZMs, ZLs) and three control groups (ZS, ZM, ZL) were designed depending on the size of the underlying zirconia particles. The test-group zirconia suspensions were prepared through silanization, which was not applied to the control-group suspensions. Depending on the particle size, viscosity differences between the test and control groups were 16,842, 18,623, and 12,303 mPa·s, respectively. Compared to the other groups, the viscosity of the ZLs group suspension decreased by 70.98–88.04%. This confirmed that the viscosity of the suspensions was affected by the particle size and the presence of silane coating. The dispersion stability of the zirconia suspensions was evaluated over 20 days. A sedimentation test confirmed that the sedimentation rate of the ZLs group was slower than those of the other groups. This study aimed to optimize the suspension manufacturing method to effectively be utilized in further commercializing zirconia three-dimensional (3D) printing and could also help to develop various medical applications.

Keywords: additive manufacture; dispersion stability; silane coupling agent; viscosity; zirconia suspension

Citation: Jun, M.-J.; Kang, J.-H.; Sakthiabirami, K.; Hosseini Toopghara, S.A.; Kim, Y.-S.; Yun, K.-D.; Park, S.-W. The Impact of Particle Size and Surface Treatment of Zirconia Suspension for Photocuring Additive Manufacturing. *Materials* **2023**, *16*, 1670. https://doi.org/10.3390/ma16041670

Academic Editors: Gianrico Spagnuolo and Matthias Kern

Received: 19 December 2022
Revised: 11 February 2023
Accepted: 14 February 2023
Published: 16 February 2023

1. Introduction

Zirconia is an oxide ceramic with excellent mechanical properties, and it can overcome the problems of low strength and low hardness of the existing ceramic prosthesis materials. For this reason, zirconia has attracted considerable interest in the field of dentistry because of its esthetic appearance with a similar color to that of human teeth [1–3].

Over the years, zirconia prostheses have been manufactured using several methods, including the cutting of pre-sintered and fully sintered zirconia by using CAD/CAM (computer-aided design/computer-aided manufacturing) or CM (copy-milling) systems [4,5]. Zirconia prostheses processed using CAD/CAM systems may have rough scratches or defects on their surfaces, which eventually develop into cracks after sintering and may reduce their strength [1,6,7]. For this reason, in the dental field, additive manufacturing-type three-dimensional (3D) printing technology has recently been introduced as a manufacturing process that provides high precision. Photopolymerization-based technology (photocuring additive manufacturing) employs a liquid solution that is solidified owing to the photocuring reaction of a photosensitive resin under ultraviolet (UV) irradiation. Successive layers are stacked on top of the last layer [8–10]. In addition, with photocuring additive manufacturing, it is possible to achieve high precision, excellent surface quality, and high printing speeds [11–16].

To ensure that the prostheses manufactured using the zirconia additive manufacturing method exhibit clinically applicable physical performance, a high content of zirconia particles (40–60 vol.%) must be included in the binder [15]. Here is where the problem arises: as the zirconia particle content increases, the viscosity, light scattering, and refractive index of the suspension increase; and the dispersion, stability, and fluidity of the particles in the suspension decrease, which adversely affect the 3D printing process and act as an obstacle for obtaining better mechanical properties. [17–25]. As the fraction of solid particles added to the binder increases, the possibility of collisions between the particles increases, and these interparticle collisions hinder particle movement in the suspension. During particle collision, friction adds additional shear force, which leads to an increase in viscosity (Figure 1a). When the solid fraction reaches its maximum value (φmax), the force required to shear the sample increases significantly owing to interparticle interactions and exceeds the required shear stress [26]. At certain solid fraction values, when the interparticle interactions intensify, shear thinning is observed. The viscosity difference between small and large particles decreases at higher shear rates because the particles are rearranged favorably in the flow direction (Figure 1b) [26]. Suspensions containing spheres of both sizes exhibit shear thinning; however, the phenomenon is more pronounced in the suspension prepared using the smaller spheres.

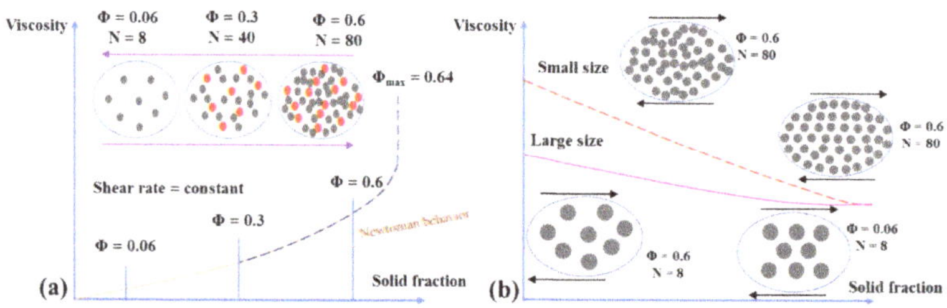

Figure 1. (**a**) The viscosity of sphere-bearing suspensions versus solid fraction under a constant shear rate and constant sphere size. (**b**) Shear-rate dependence of small and large spheres for a continuous solid fraction (Grey spheres indicate the particle, whereas red spheres denote the particle collision which occurs due to increased solid fraction and arrows indicate the shear stress).

To improve the behavior of and solve the problems associated with zirconia suspensions that contain a high fraction of solid particles, studies are being conducted to achieve low viscosity while maintaining a high particle content by subjecting the particles to surface treatment using a silane coupling agent [27–33]. In general, a large number of hydroxyl groups are present on the surface of ceramic particles; and, as a result, these particles have hydrophilic properties, and they tend to aggregate [34]. The silane coupling agent transforms the hydrophilic surface of ceramic particles into a hydrophobic surface. The hydrophobic anchoring head group and hydrophilic end chain create a steric barrier that ensures uniform dispersion of particles in suspensions [35]. For this reason, silane coupling agents modify the surface of ceramic particles and increase their dispersion stability [20], improve their interfacial adhesion, and facilitate the preparation of high-volume-fraction suspensions. The aim of the present study is to optimize the suspension's poor rheological properties through identifying the factors that affect zirconia suspensions with a silane coupling agent in various particle sizes for possible utilization in further photocurable additive manufacturing or other techniques. In addition, the poor rheological properties of the high-content zirconia suspensions were enhanced, which would positively affect the 3D printing process.

2. Experimental Details

2.1. Experimental Material

In this experiment, three types of zirconia powders (TZ-3Y, TZ-3YS-E, Tosho, Anjo Shi, Japan; TZY-3, Qingdao Terio Corporation, Qingdao, China) were used to prepare a zirconia suspension (Table 1). Methyltrimethoxysilane (MTMES, SAMCHUN Chemical, Seoul, Korea) was used as a silane coupling agent to modify the surfaces of the zirconia powders. The suspension was prepared using a UV-curable resin as the base, which was a mixture of 1,6-Hexanediol diacrylate (HDDA, Sigma Aldrich Inc., St. Louis, MO, USA), photocurable resin (XYZPRINTING rigid, New Taipei City, Taiwan), photoinitiator (Irgacure 819, Ciba Specialty Chemicals, Basel, Switzerland), and a dispersant called Disperbyk (BYK-180, BYK, Chester, NY, USA).

Table 1. Zirconia particle sizes considered in this study.

Powder	Code	Commercial Size (nm)
TZ-3Y	ZS	40
TZ-3YS-E	ZM	90
TZY-3	ZL	300–600

2.2. BET Analysis

In BET analysis, the surface of the samples meets a gas at a series of pressures that will absorb the surface and the walls of the pores in the samples. Various gases could be used; however, nitrogen is the most common absorbate. The reason is that N_2 is a diatomic non-spherical, and it has a quadrupole moment. This will lead it to preferentially absorb to the surface and remain for longer. So, the surface area and pore volume of the powders were analyzed by conducting N_2 adsorption/desorption measurement at 77.3 K by using a surface area and pore-size analyzer (BELSORP-mini II, Microtracbel, Osaka, Japan).

2.3. Particle Size Analysis and Zeta Potential Measurement

We used DLS)Zetasizer NanoZS90, Malvern Instrument Ltd., Worcestershire, UK) to measure and compare the particle size and surface charge of the zirconia particles. Dynamic Light Scattering (DLS), is a precise, non-invasive, fast, reliable, technology that is well established for size measurement and study of the size distribution of particles in submicron sizes. Samples were dispersed in ethanol and 1 mL of each sample was transferred to a quartz glass cuvette. The measurements were performed thrice, and the averages of the three values were used in this study.

2.4. Silane-Coated Zirconia Powder Manufacturing

To apply silane coating on the selected zirconia particles (ZS, ZM, ZL), distilled water and ethyl alcohol were mixed in a ratio of 20:1. Acetic acid (64-19-7, Daejung Chemicals Co., Ltd., Busan, Republic of Korea) was added and stirred into this solution until the solution pH was 4, as measured using a pH meter (OROION STAR A21, ThermoFisher, Waltham, MA, USA). MTMES (2 mL) was added to the prepared pH 4 solution and stirred at 40 rpm for 60 min at 50 °C to form silanol. Approximately 25 g of each of the three zirconia powders of various particle sizes was added to 50 mL of the silanol solution and stirred at 500 rpm for 25 min to achieve homogeneous mixing. Then, the solutions were dried for 24 h at 80 °C in a drying oven (p-6 Planetary Mill, Fritsch, Co., Ltd., Idar-Oberstein, Germany). Thereafter, the dried zirconia powders were classified using a 100 μm sieve, and the final silane-coated zirconia powders (ZSs, ZMs, ZLs) were obtained. This experiment was subdivided into six groups: three control groups (ZS, ZM, ZL) and three experimental groups (ZSs, ZMs, ZLs). A flowchart of powder modification with silane is presented in Figure 2.

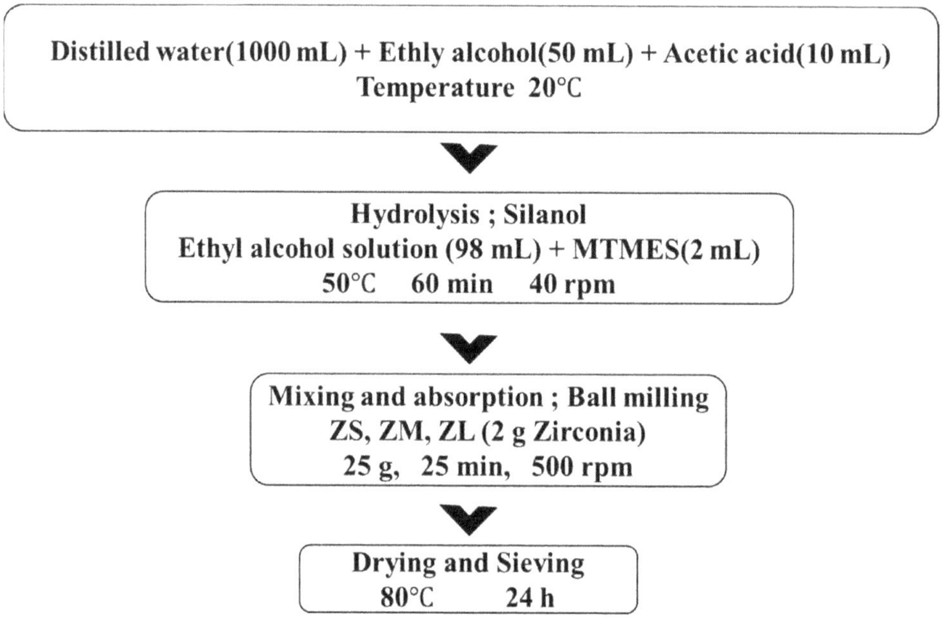

Figure 2. Flowchart of powder modification and suspension.

2.5. *Fourier-Transform Infrared Spectroscopy (FT-IR)*

This technique uses IR radiation to determine the chemical structures of the samples. When the radiation passes through the sample, some radiation will absorb and some of them will be transmitted. This spectrum acts like a fingerprint to determine the chemical structures. The chemical structures of the zirconia particles included in the control and experimental groups and the presence of silane coating on them were determined in the range of 450–5000 cm^{-1} using Fourier-transform infrared spectroscopy (FTIR, Spectrum 400, Perkin Elmer, Waltham, MA, USA).

2.6. *Observation of Surface-Treated Zirconia*

We used a transmission electron microscope (field-emission transmission electron microscope, JEM-2100F, JEOL Ltd., Akishima, Japan) to observe and compare the coating morphologies of MTMES on the surfaces of the nanoparticles in the experimental and control groups.

2.7. *Slurry Preparation*

Zirconia photocuring suspensions were prepared with a volume fraction of 40 vol.% by using zirconia powders from the three experimental and three control groups. MTMES was added to suspensions belonging to the experimental groups; and HDDA, resin, and photoinitiator were used as binders in a ratio of 7:3:0.1 for mixing all the powders. In the end, Disperbyk (BYK-180, BYK, USA) was added to all the suspensions. To ensure the homogeneous mixing of the suspensions, they were stirred in a vacuum for 200 s by using a planetary centrifugal mixer (ARV-310, Thinky Corp., Tokyo, Japan). The compositions of the prepared solutions are summarized in Table 2.

Table 2. Experimental groups used in this study (vol.%).

Group (40 vol.%)		Zirconia	HDDA	Resin	Silane	Photo Initiator	Disperbyk	Total wt. (vol.%)
Control	ZS	40.00	39.54	15.47	0	0.50	4.48	100
	ZM	40.00	39.54	15.47	0	0.50	4.48	100
	ZL	40.00	39.54	15.47	0	0.50	4.48	100
Experiment	ZSs	40.00	35.93	14.06	5.07	0.45	4.48	100
	ZMs	40.00	35.93	14.06	5.07	0.45	4.48	100
	ZLs	40.00	35.93	14.06	5.07	0.45	4.48	100

2.8. Observation of Colloidal Stability

The colloidal stability of the prepared zirconia suspensions was visually observed to determine their stability and the effect of the added dispersant. To this end, suspensions containing less than 1 vol.% of zirconia particles were used. Approximately, 10 mL of each suspension was poured into separate glass vials, and their stability was observed for up to 2 weeks (0, 5, 10, 15, and 20 days) until a clear sign was observed.

2.9. Viscosity

The viscosity values of the suspensions belonging to each group were measured using a viscometer (DV3T, Brookfield Engineering Laboratories, Stoughton, MA, USA). For this goal, various types of spindles exist and they were used based on the sample viscosities. The disk spindles produce accurate and reproducible results in the viscosity range of our suspensions.

To this end, the suspensions were placed at the center of a circular plate, and the distance between the plate and the spindle (CP4005, Brookfield Engineering Laboratories, Stoughton, MA, USA) was set to 2 mm. The viscosity was measured by changing the rotational speed of the cone at a constant temperature. Each group was measured at a shear rate of 1.2 to 20/s at 25 °C.

3. Results

3.1. Brunauer–Emmett–Teller Analysis

Brunauer–Emmett–Teller (BET) analysis was performed to determine the pore size and specific surface area of the selected zirconia particles (Table 3). Among the selected zirconia particles, the measured specific surface area and average pore size of ZS were 15.344 m^2g^{-1} and 58.3 nm, respectively; those of ZM were 9.3932 m^2g^{-1} and 31 nm; and those of ZL were 11.4 m^2g^{-1} and 34.4 nm. These results confirmed that the total pore volume differed depending on the average pore size, which affected the specific surface area as well.

Table 3. BET results of different powders.

Powder (Code)	Surface Area [m^2/g]	Total Pore Volume (p/p^0 = 0.990)	Mean Pore Diameter [nm]
TZ-3Y (ZS)	15.3	0.22	58.3
TZ-3YS-E (ZM)	9.4	0.05	31
TZY-3 (ZL)	11.4	0.1	33.4

3.2. Particle Size Analysis and Zeta Analysis

Through dynamic light scattering, we measure and compare the particle size and surface charge of the zirconia particles. The respective results are demonstrated in Figures 3 and 4.

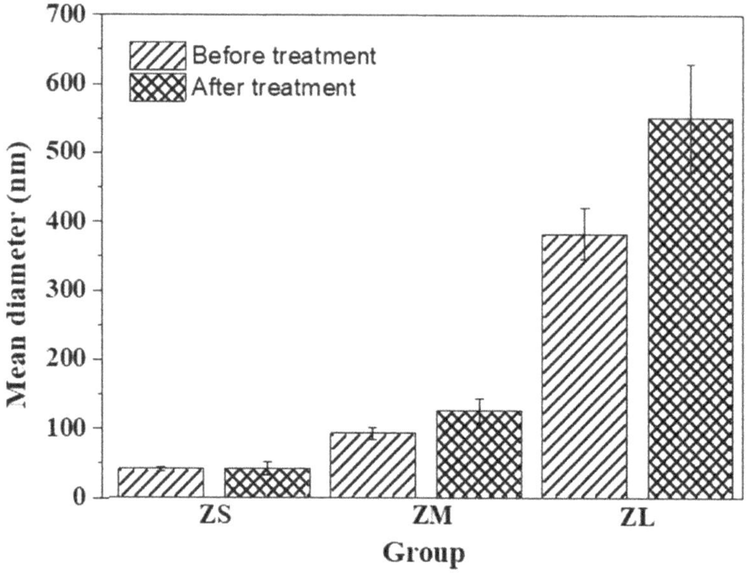

Figure 3. Size distribution of zirconia particles before and after silane coating.

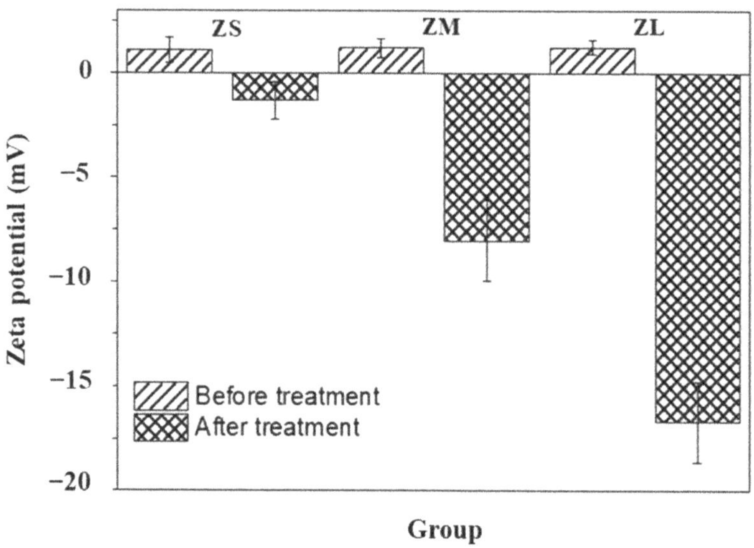

Figure 4. Zeta potentials of zirconia particles before and after silane coating.

3.3. Surface Treatment Analysis of Zirconia Particles

FE-TEM was used to identify the presence or absence of silanol coating on the control and experimental group particles, and depicted in the following Figures 5 and 6.

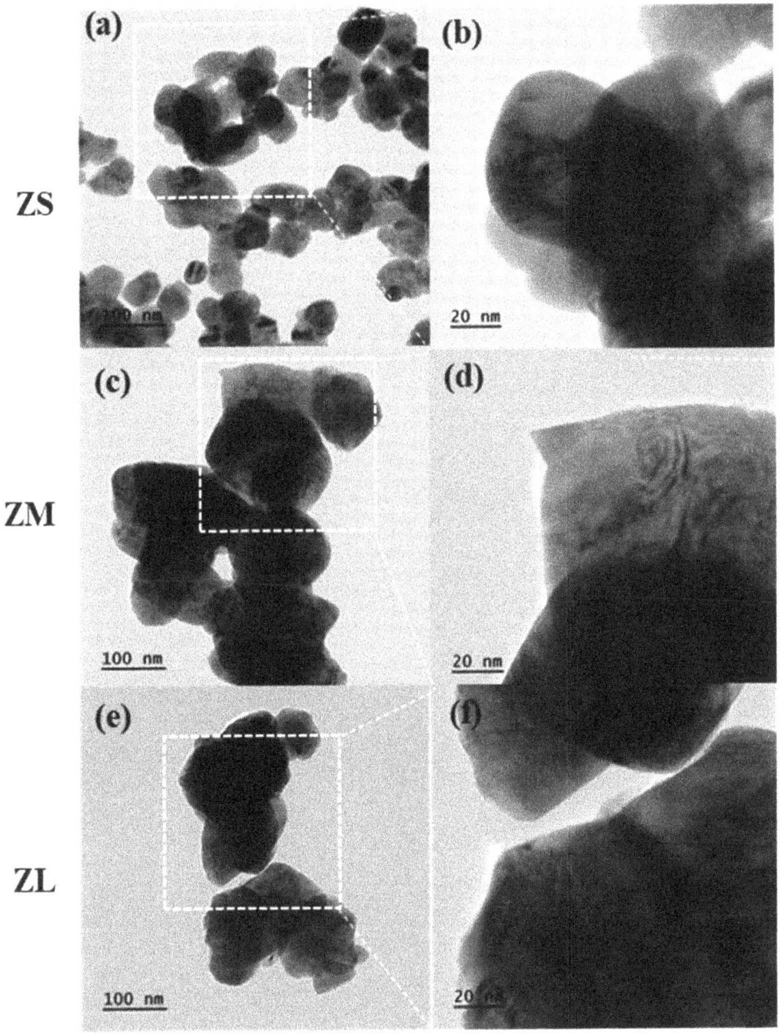

Figure 5. TEM images of control groups (ZS, ZM, ZL): (**a**) ZS ×100K, (**b**) ZS ×200K, (**c**) ZM ×100K, (**d**) ZM ×200K, (**e**) ZL ×100K, and (**f**) ZL ×200K.

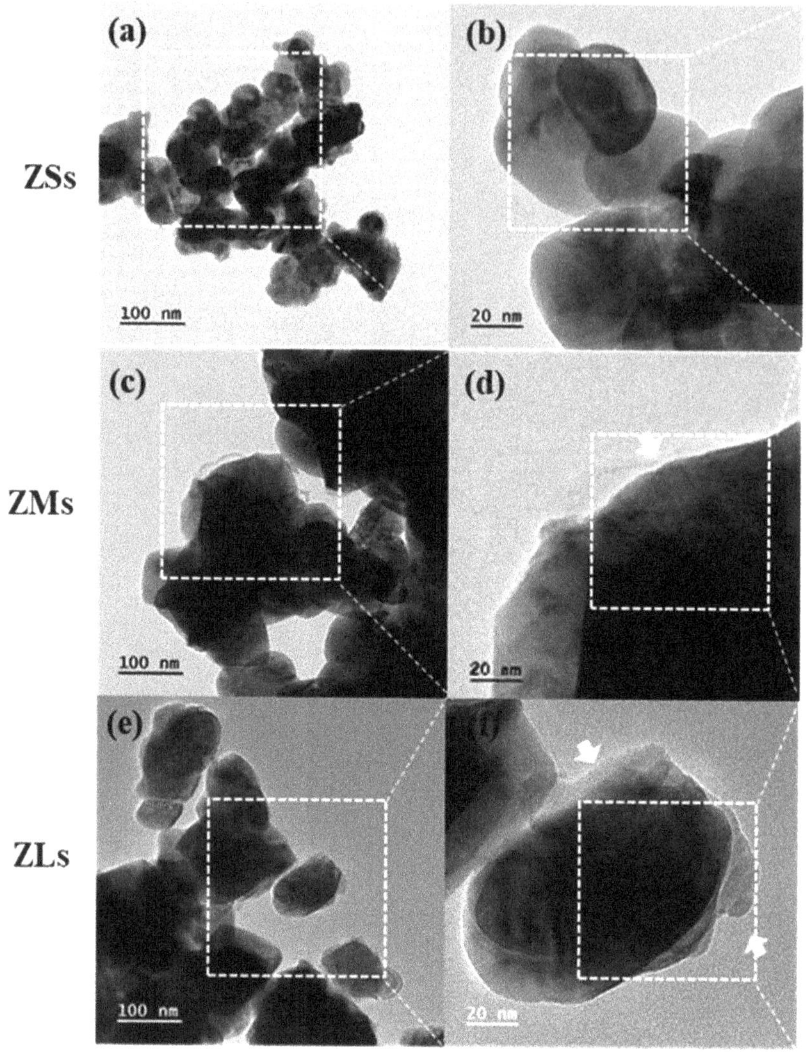

Figure 6. TEM images of experimental groups (ZSs, ZMs, ZLs): (**a**) ZSs ×100K, (**b**) ZSs ×200K, (**c**) ZMs ×100K, (**d**) ZMs ×200K, (**e**) ZLs ×100K, and (**f**) ZLs ×200K. Arrow indicate silane coating on the zirconia particle.

3.4. *FT-IR Analysis*

To prove the effect of silane coating on the zirconia particles, the chemical reactions of the powders belonging to the control and experimental groups were analyzed by means of FT-IR spectroscopy (Figure 7).

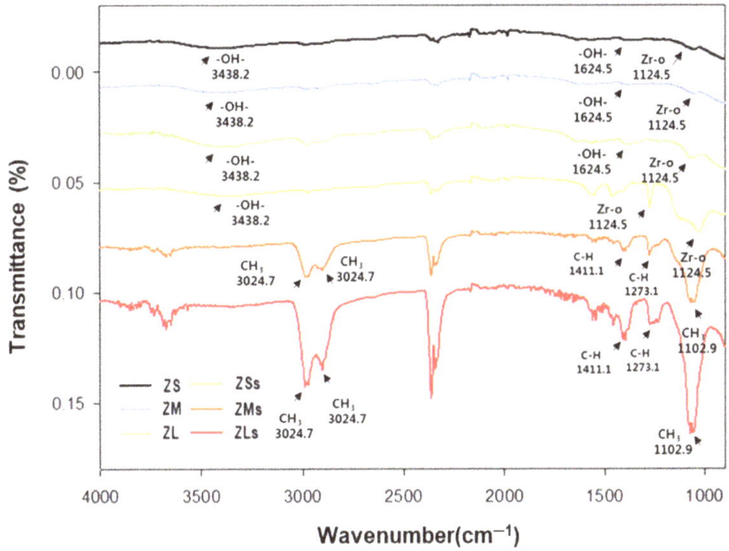

Figure 7. FT-IR spectra of various zirconia particles before and after silane coating.

3.5. Stability and Sedimentation Evaluation of Zirconia Suspensions

In the sedimentation test of the prepared zirconia suspensions belonging to both groups, they were visually evaluated from day 0 to day 20 to confirm their stability [5]. It was confirmed that the zirconia particles precipitated to different degrees over the set periods of 5, 10, 15, and 20 days. After 10 days, rapid particle precipitation was observed in the control groups.

Moreover, rapid precipitation was observed in the experimental ZSs group. After 20 days, ZMs and ZLs remained as suspensions, and their dispersibility was maintained, as could be observed with the naked eye (Figure 8).

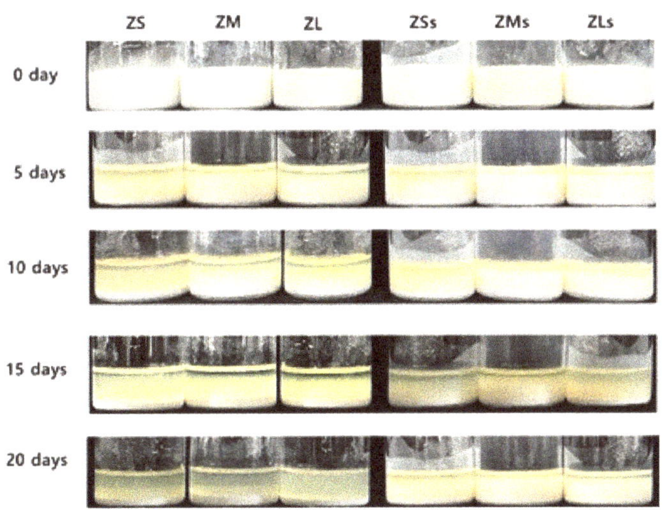

Figure 8. Sedimentation picture of suspensions with increasing days of before and after silane coating.

3.6. Viscosity

The viscosity of the zirconia suspensions belonging to the control and experimental groups were evaluated; and the rheological behaviors of the suspensions depending on the zirconia particle size, and the presence or absence of silane coating are depicted in Figure 9.

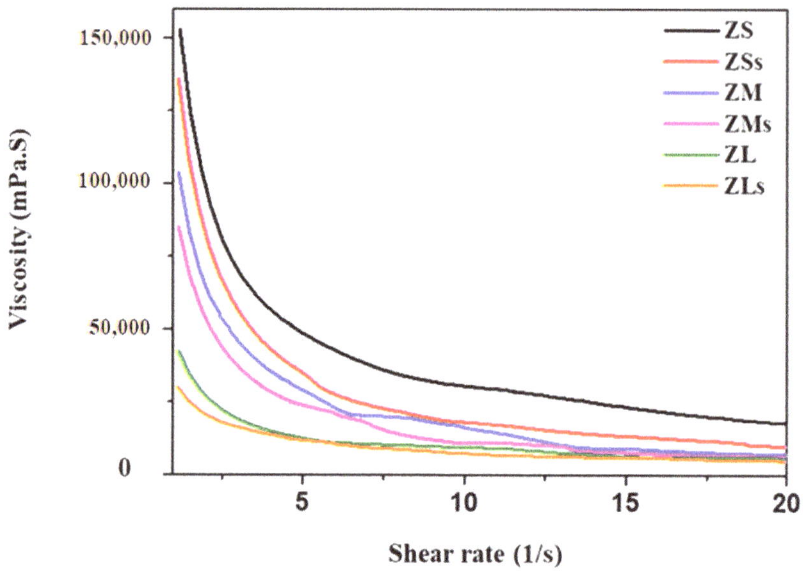

Figure 9. The viscosity of various zirconia suspensions before and after silane coating.

4. Discussion

The results of the Brunauer–Emmett–Teller (BET) analysis confirmed that the total pore volume differed depending on the average pore size, which affected the specific surface area as well. Particle size analysis was performed to confirm the change in particle size depending on whether the particles were subjected to surface treatment. In the case of the experimental groups (ZSs, ZMs, ZLs: 18.49 $\pm$ 0.14, 268.66 $\pm$ 17.35, 570.06 $\pm$ 570.24 nm, respectively), an increase in particle size was confirmed. In particular, the particle size of the ZLs group was the largest (Figure 3). By contrast, while ZSs had the smallest particle size among the experimental groups treated with silane, its particle size did not change significantly post-treatment. Owing to the high specific surface areas of ZM and ZL, their surfaces may not have been adequately coated with silane regarding the limited amount of it. It was confirmed that the organic coating of particles could be affected by various factors, such as particle size, pore size, pore area, and the amount of organic material used for coating. These factors should be adjusted and optimized depending on the characteristics of the particles [23,26,36].

The zeta potential measurements confirmed that the control groups exhibited weak anodic characteristics, while the zeta potential of the experimental groups changed to cathodic after silane treatment. This electrode reversal occurred because of the silanol treatment, and it was confirmed that the zeta potential value increased as the size of the particles in the experimental groups increased. The higher the absolute value of the repulsive electric force of a colloid suspended in a liquid, the higher the potential difference in the diffusion layer, which affects the uniform dispersion of the particles [37,38]. The ZLs group had the highest zeta potential value, and its viscosity and suspension stability was expected to improve because of the silanol treatment.

In the case of the ZSs group, the presence of a silane coating could not be confirmed. By contrast, in cases of the ZMs and ZLs groups, the presence of a silane coating was confirmed.

In addition, we were able to confirm that the coating layer on the ZLs group particles was thicker because the number of particles and their specific surface area increased as the particle size decreased, even though the same quantities of zirconia particles and silane coupling agent were used across groups. The capacity of the limited silane coupling agent leads to differences in the coating thickness of the particles depending on the coating effect under various conditions (e.g., size, surface area, surface environment) [24,26].

To ensure effective silane coating of zirconia particles, it is necessary to determine the appropriate ratio of particle size to coating agent.

Among the results of the FR-IR spectroscopy of the silane-coated experimental groups, stretching vibrations of the characteristic C-H bands of silane appeared close to 1411.1 cm^{-1}, 1273.1 cm^{-1}, and 1102.9 cm^{-1} in the spectra of ZMs and ZLs. Moreover, –CH3 oscillations were observed at 3024.7 cm^{-1} and 2841.2 cm^{-1} [23,39]. Among the control groups, which were not coated with silane, only ZrO_2-derived –OH bands at 3438.2 cm^{-1} and 1624 cm^{-1} and a Zr-O-derived band at 1124.5 cm^{-1} were observed [11]. The peaks at 1411.1 cm^{-1} and 1273.1 cm^{-1} were attributed to CH stretching (CH=CH2) due to the silane coating [40,41]. These results demonstrated that the silane coupling agent was coated on the zirconia particles.

The sedimentation test of the prepared zirconia suspensions confirmed that the zirconia particles precipitated to different degrees over the set periods of 5, 10, 15, and 20 days. After 10 days, rapid particle precipitation was observed in the control groups.

Moreover, rapid precipitation was observed in the experimental ZSs group. After 20 days, ZMs and ZLs remained as suspensions, and their dispersibility was maintained, as could be observed with the naked eye (Figure 8). This result could possibly be ascribed to the surface treatment of the ceramic particles, which increased their affinity toward the organic system of the photopolymerizable binder, and helped to maintain excellent particle dispersion and stability [42].

All the zirconia suspensions, prepared using a zirconia volume fraction of 40 vol.%, exhibited non-Newtonian shear-thinning behavior under shear rates of 1.2–20/s^{-1} (Figure 8). It was confirmed that the viscosity of these suspensions increased as the size of the added zirconia particles decreased. Compared to the viscosities of ZSs and ZLs, the viscosities of these suspensions decreased by 72.62–80.28% owing to the applied shear rates.

The viscosity values of the suspensions belonging to the experimental group were lower than those of the suspensions belong to the control group (Table 4).

Table 4. Change in viscosity with shear rate.

Shear Rate	ZS (mPa·s)	ZSs (mPa·s)	ZM (mPa·s)	ZMs (mPa·s)	ZL (mPa·s)	ZLs (mPa·s)
20 s^{-1}	18,026	10,015	7047	6418	5789	4935
1.2 s^{-1}	152,685	135,843	103,697	85,074	42,407	30,104

This result was ascribed to the fact that the powders belonging to the experimental group reacted more effectively than those belonging to the control group and were better dispersed in the binder containing a dispersant that was involved in interparticle electrostatic interactions [43]. To realize uniform layer formation in the products manufactured using digital light processing (DLP), an additive manufacturing process, the suspension must have the appropriate viscosity characteristics and shear-thinning behavior to ensure that the photocuring mechanism and fluidity are maintained [44].

5. Conclusions

This study demonstrates the influence of zirconia particle size and the surface treatment using a silane coupling agent for the preparation of a suspension with optimal rheological properties. Based on the results, it is significant that the characteristics of candidate particles should be adequately analyzed before selecting them for use to ensure effective surface treatment and 3D additive manufacturing. In addition, a zirconia suspension with excellent features can be prepared by determining the appropriate ratio of coating agent suitable for the characteristics of the particles at hand. Although our experiments have a few limitations, for instance, different ratios of silane-coating agents were not used, we believe that the results of this study will act as a potential candidate in the field of ceramic additive manufacturing.

Author Contributions: Conceptualization and writing—original draft preparation, S.-W.P., J.-H.K. and M.-J.J.; writing—review and editing, S.-W.P., J.-H.K., K.S. and K.-D.Y.; investigation and data curation, M.-J.J., J.-H.K. and S.-W.P.; methodology, M.-J.J., Y.-S.K. and S.A.H.T.; visualization, S.A.H.T.; supervision, S.-W.P. and K.-D.Y.; funding acquisition, J.-H.K. and S.-W.P. All authors have read and agreed to the published version of the manuscript.

Funding: This work was supported by a National Research Foundation of Korea Grant funded by the Korean government (MSIP) (Grant nos. 2022R1A2C2009633 and 2022R1A6A3A01087006).

Institutional Review Board Statement: Not applicable.

Informed Consent Statement: Not applicable.

Data Availability Statement: All the data have been illustrated in the manuscript.

Conflicts of Interest: The authors declare no conflict of interest.

References

1. Denry, I.; Kelly, J.R. State of the art of zirconia for dental applications. *Dent. Mater.* **2008**, *24*, 299–307. [CrossRef] [PubMed]
2. Schriwer, C.; Skjold, A.; Gjerdet, N.R.; Øilo, M. Monolithic zirconia dental crowns. Internal fit, margin quality, fracture mode and load at fracture. *Dent. Mater.* **2017**, *33*, 1012–1020. [CrossRef] [PubMed]
3. Shareef, M.; Van Noort, R.; Messer, P.; Piddock, V. The effect of microstructural features on the biaxial flexural strength of leucite reinforced glass-ceramics. *J. Mater. Sci. Mater. Med.* **1994**, *5*, 113–118. [CrossRef]
4. Sundh, A.; Sjögren, G. A study of the bending resistance of implant-supported reinforced alumina and machined zirconia abutments and copies. *Dent. Mater.* **2008**, *24*, 611–617. [CrossRef]
5. Wang, H.; Aboushelib, M.N.; Feilzer, A.J. Strength influencing variables on CAD/CAM zirconia frameworks. *Dent. Mater.* **2008**, *24*, 633–638. [CrossRef]
6. Preis, V.; Behr, M.; Handel, G.; Schneider-Feyrer, S.; Hahnel, S.; Rosentritt, M. Wear performance of dental ceramics after grinding and polishing treatments. *J. Mech. Behav. Biomed. Mater.* **2012**, *10*, 13–22. [CrossRef]
7. Preis, V.; Behr, M.; Hahnel, S.; Handel, G.; Rosentritt, M. In vitro failure and fracture resistance of veneered and full-contour zirconia restorations. *J. Dent.* **2012**, *40*, 921–928. [CrossRef]
8. He, R.; Liu, W.; Wu, Z.; An, D.; Huang, M.; Wu, H.; Jiang, Q.; Ji, X.; Wu, S.; Xie, Z. Fabrication of complex-shaped zirconia ceramic parts via a DLP-stereolithography-based 3D printing method. *Ceram. Int.* **2018**, *44*, 3412–3416. [CrossRef]
9. Halloran, J.W.; Tomeckova, V.; Gentry, S.; Das, S.; Cilino, P.; Yuan, D.; Guo, R.; Rudraraju, A.; Shao, P.; Wu, T. Photopolymerization of powder suspensions for shaping ceramics. *J. Eur. Ceram. Soc.* **2011**, *31*, 2613–2619. [CrossRef]
10. Mitteramskogler, G.; Gmeiner, R.; Felzmann, R.; Gruber, S.; Hofstetter, C.; Stampfl, J.; Ebert, J.; Wachter, W.; Laubersheimer, J. Light curing strategies for lithography-based additive manufacturing of customized ceramics. *Addit. Manuf.* **2014**, *1*, 110–118. [CrossRef]
11. Hinczewski, C.; Corbel, S.; Chartier, T. Ceramic suspensions suitable for stereolithography. *J. Eur. Ceram. Soc.* **1998**, *18*, 583–590. [CrossRef]
12. Chartier, T.; Chaput, C.; Doreau, F.; Loiseau, M. Stereolithography of structural complex ceramic parts. *J. Mater. Sci.* **2002**, *37*, 3141–3147. [CrossRef]
13. Schwentenwein, M.; Schneider, P.; Homa, J. Lithography-based ceramic manufacturing: A novel technique for additive manufacturing of high-performance ceramics. *Adv. Sci. Technol.* **2014**, *88*, 60–64.
14. Song, X.; Chen, Y.; Lee, T.W.; Wu, S.; Cheng, L. Ceramic fabrication using Mask-Image-Projection-based Stereolithography integrated with tape-casting. *J. Manuf. Process.* **2015**, *20*, 456–464. [CrossRef]
15. Wu, H.; Cheng, Y.; Liu, W.; He, R.; Zhou, M.; Wu, S.; Song, X.; Chen, Y. Effect of the particle size and the debinding process on the density of alumina ceramics fabricated by 3D printing based on stereolithography. *Ceram. Int.* **2016**, *42*, 17290–17294. [CrossRef]
16. Halloran, J.W. Ceramic stereolithography: Additive manufacturing for ceramics by photopolymerization. *Annu. Rev. Mater. Res.* **2016**, *46*, 19–40. [CrossRef]
17. Griffith, M.L.; Halloran, J.W. Freeform fabrication of ceramics via stereolithography. *J. Am. Ceram. Soc.* **1996**, *79*, 2601–2608. [CrossRef]
18. Tay, B.; Evans, J.; Edirisinghe, M. Solid freeform fabrication of ceramics. *Int. Mater. Rev.* **2003**, *48*, 341–370. [CrossRef]

19. Kambly, K. *Characterization of Curing Kinetics and Polymerization Shrinkage in Ceramic-Loaded Photocurable Resins for Large Area Maskless Photopolymerization (LAMP)*; Georgia Institute of Technology: Atlanta, GA, USA, 2009.
20. Bae, C.J.; Halloran, J.W. Integrally cored ceramic mold fabricated by ceramic stereolithography. *Int. J. Appl. Ceram. Technol.* **2011**, *8*, 1255–1262. [CrossRef]
21. Song, S.Y.; Park, M.S.; Lee, D.; Lee, J.W.; Yun, J.S. Optimization and characterization of high-viscosity ZrO_2 ceramic nanocomposite resins for supportless stereolithography. *Mater. Des.* **2019**, *180*, 107960. [CrossRef]
22. Song, S.Y.; Park, M.S.; Lee, J.W.; Yun, J.S. A study on the rheological and mechanical properties of photo-curable ceramic/polymer composites with different silane coupling agents for SLA 3D printing technology. *Nanomaterials* **2018**, *8*, 93. [CrossRef] [PubMed]
23. Song, S.Y.; Park, M.S.; Lee, J.W.; Yun, J.S. Improvement of dispersion stability and 3D-printing characteristics of ceramics in photopolymers by controlling the coating thickness of silane coupling agents. *Mater. Chem. Phys.* **2018**, *216*, 446–453. [CrossRef]
24. Melcher, R.; Martins, S.; Travitzky, N.; Greil, P. Fabrication of Al_2O_3-based composites by indirect 3D-printing. *Mater. Lett.* **2006**, *60*, 572–575. [CrossRef]
25. Licciulli, A.; Corcione, C.E.; Greco, A.; Amicarelli, V.; Maffezzoli, A. Laser stereolithography of ZrO_2 toughened Al_2O_3. *J. Eur. Ceram. Soc.* **2005**, *25*, 1581–1589. [CrossRef]
26. Guazzelli, É.; Pouliquen, O. Rheology of dense granular suspensions. *J. Fluid Mech.* **2018**, *852*, 1. [CrossRef]
27. Halloran, J.W.; Griffith, M.; Chu, T.-M. Stereolithography Resin for Rapid Prototyping of Ceramics and Metals. U.S. Patent US6117612A, 12 September 2000.
28. Ebert, J.; Özkol, E.; Zeichner, A.; Uibel, K.; Weiss, Ö.; Koops, U.; Telle, R.; Fischer, H. Direct inkjet printing of dental prostheses made of zirconia. *J. Dent. Res.* **2009**, *88*, 673–676. [CrossRef] [PubMed]
29. Kim, M.; Song, H.Y.; Choi, W.J.; Hyun, K. Evaluation of the degree of dispersion of polymer nanocomposites (PNCs) using nonlinear rheological properties by FT-rheology. *Macromolecules* **2019**, *52*, 8604–8616. [CrossRef]
30. Di Maro, M.; Duraccio, D.; Malucelli, G.; Faga, M. High density polyethylene composites containing alumina-toughened zirconia particles: Mechanical and tribological behavior. *Compos. Part B Eng.* **2021**, *217*, 108892. [CrossRef]
31. Lee, S.; min Kim, H.; Seong, D.G.; Lee, D. Synergistic improvement of flame retardant properties of expandable graphite and multi-walled carbon nanotube reinforced intumescent polyketone nanocomposites. *Carbon* **2019**, *143*, 650–659. [CrossRef]
32. Cho, J.; Joshi, M.; Sun, C. Effect of inclusion size on mechanical properties of polymeric composites with micro and nano particles. *Compos. Sci. Technol.* **2006**, *66*, 1941–1952. [CrossRef]
33. Kim, D.H.; Fasulo, P.D.; Rodgers, W.R.; Paul, D. Effect of the ratio of maleated polypropylene to organoclay on the structure and properties of TPO-based nanocomposites. Part I: Morphology and mechanical properties. *Polymer* **2007**, *48*, 5960–5978. [CrossRef]
34. Lertwimolnun, W.; Vergnes, B. Influence of compatibilizer and processing conditions on the dispersion of nanoclay in a polypropylene matrix. *Polymer* **2005**, *46*, 3462–3471. [CrossRef]
35. Lee, T.W.; Lee, S.; Park, S.-M.; Lee, D. Mechanical, thermomechanical, and local anisotropy analyses of long basalt fiber reinforced polyamide 6 composites. *Compos. Struct.* **2019**, *222*, 110917. [CrossRef]
36. Shinkaryov, A.S.; Cherkasova, M.V.; Pelevin, I.A.; Ozherelkov, D.Y.; Chernyshikhin, S.V.; Kharitonova, N.A.; Gromov, A.A.; Nalivaiko, A.Y. Aluminum Powder Preparation for Additive Manufacturing Using Electrostatic Classification. *Coatings* **2021**, *11*, 629. [CrossRef]
37. Park, E.; Bac, L.; Kim, J.; Kwon, Y.; Kim, J.; Choi, H.; Chung, Y. Production and properties of Ag metallic nanoparticle fluid by electrical explosion of wire in liquid. *J. Powder Mater.* **2009**, *16*, 217–222. [CrossRef]
38. Liu, Y.; Cheng, L.; Li, H.; Li, Q.; Shi, Y.; Liu, F.; Wu, Q.; Liu, S. Formation mechanism of stereolithography of Si_3N_4 slurry using silane coupling agent as modifier and dispersant. *Ceram. Int.* **2020**, *46*, 14583–14590. [CrossRef]
39. Goswami, A.; Ankit, K.; Balashanmugam, N.; Umarji, A.M.; Madras, G. Optimization of rheological properties of photopolymerizable alumina suspensions for ceramic microstereolithography. *Ceram. Int.* **2014**, *40*, 3655–3665. [CrossRef]
40. Jang, K.-J.; Kang, J.-H.; Fisher, J.G.; Park, S.-W. Effect of the volume fraction of zirconia suspensions on the microstructure and physical properties of products produced by additive manufacturing. *Dent. Mater.* **2019**, *35*, e97–e106.
41. Gentry, S.P.; Halloran, J.W. Light scattering in absorbing ceramic suspensions: Effect on the width and depth of photopolymerized features. *J. Eur. Ceram. Soc.* **2015**, *35*, 1895–1904. [CrossRef]
42. Sharma, L.; Kumar, M.; Saxena, A.; Chand, M.; Gupta, J. Influence of pore size distribution on Pt dispersion in Pt-Sn/Al_2O_3 reforming catalyst. *J. Mol. Catal. A Chem.* **2002**, *185*, 135–141. [CrossRef]
43. Griffith, M.L.; Halloran, J.W. Scattering of ultraviolet radiation in turbid suspensions. *J. Appl. Phys.* **1997**, *81*, 2538–2546. [CrossRef]
44. Tesavibul, P.; Chantaweroad, S.; Laohaprapanon, A.; Channasanon, S.; Uppanan, P.; Tanodekaew, S.; Chalermkarnnon, P.; Sitthiseripratip, K. Biocompatibility of hydroxyapatite scaffolds processed by lithography-based additive manufacturing. *Bio-Med. Mater. Eng.* **2015**, *26*, 31–38. [CrossRef] [PubMed]

materials

MDPI

Article

Effects of UV Absorber on Zirconia Fabricated with Digital Light Processing Additive Manufacturing

Jin-Ho Kang [1,†], Kumaresan Sakthiabirami [1,2,†], Hyun-Ah Kim [1], Seyed Aliakbar Hosseini Toopghara [3], Mee-Jin Jun [4], Hyun-Pil Lim [1], Chan Park [1], Kwi-Dug Yun [1,*] and Sang-Won Park [1,2,*]

1 Department of Prosthodontics, School of Dentistry, Chonnam National University, Gwangju 61186, Republic of Korea
2 Biomedical Evaluation and Research Centre, School of Dentistry, Chonnam National University, Gwangju 61186, Republic of Korea
3 Department of Medical Engineering Joint Research, Chonnam National University, Gwangju 61186, Republic of Korea
4 Department of Dental Hygiene, Gwangju Health University, Gwangju 62287, Republic of Korea
* Correspondence: ykd@chonnam.ac.kr (K.-D.Y.); psw320@chonnam.ac.kr (S.-W.P.); Tel.: +82-62-530-5842 (K.-D.Y.); +82-62-530-5631 (S.-W.P.)
† These authors contributed equally to this work.

Abstract: This study evaluated the effect of UV absorbers on the dimensional accuracy of zirconia specimens fabricated by additive manufacturing using a digital light process. Zirconia suspension for additive manufacturing was prepared by setting the volume fractions (0, 0.005, 0.05, and 0.1%) of various UV absorbers. The effect of UV absorber content was evaluated through curing thickness, geometric overgrowth model design, linear deviation, and microstructure evaluation before and after sintering. Statistical analysis was performed by Kruskal–Wallis H and post-tested by the Bonferroni correction method. There was no significant difference in the cure depth according to the presence or absence of the UV absorber, the difference in geometric overgrowth was from 2.1 to 12.5%, and the overgrowth significantly decreased as the amount of added UV absorber increased. This result may contribute to improved precision of 3D multilayer ceramic products.

Keywords: UV absorbers; zirconia; additive manufacturing; cure depth; geometric overgrowth

Citation: Kang, J.-H.; Sakthiabirami, K.; Kim, H.-A.; Hosseini Toopghara, S.A.; Jun, M.-J.; Lim, H.-P.; Park, C.; Yun, K.-D.; Park, S.-W. Effects of UV Absorber on Zirconia Fabricated with Digital Light Processing Additive Manufacturing. *Materials* **2022**, *15*, 8726. https://doi.org/10.3390/ma15248726

Academic Editors: Csaba Balázsi and Emilio Jiménez-Piqué

Received: 25 October 2022
Accepted: 5 December 2022
Published: 7 December 2022

1. Introduction

With the development of digital dentistry, computer-aided design/computer-aided manufacturing (CAD/CAM) system zirconia prostheses have gained attention for satisfactory esthetics, high biocompatibility, and improved mechanical properties compared to existing dental ceramics [1].

Zirconia ceramic also has a variety of different applications other than dental implants and frameworks [2–4], such as electrolyte [5,6] and monolithic support for solid oxide fuel cells [7], as a part of cutting tools and blades [8], for elaborate molds [9] and other precision components in thermal and mechanical applications [10].

Currently, zirconia prostheses are fabricated by subtractive manufacturing using a CAD/CAM system. A completely sintered zirconia prosthesis is obtained by milling and sintering a pre-sintered zirconia block. Studies show that 15–30% (approximately 20%) of linear shrinkage occurs due to sintering [11]. However, the exact shrinkage rate is compensated for by setting the enlargement rate according to the manufacturer's instructions for the zirconia block. Prosthesis fabrication through subtractive manufacturing has certain disadvantages, including the consumption of materials from milling burs or remaining blocks, the possibility of microcracking due to surface roughness or defects, and difficulty reproducing complex structures [12–15].

Three-dimensional printing techniques (or additive manufacturing) have gained popularity in various fields, such as temporary prostheses, splints, and model manufacturing, due to the ability to consume less material than subtractive manufacturing and reproduce complex structures [16–18]. With the recent development of additive manufacturing, studies have aimed to apply these techniques to prosthetic manufacturing in dentistry. Studies using various 3D printers, such as photopolymerization digital light processing (DLP), stereolithography apparatus (SLA), selective laser sintering (SLS), and spraying polyjet printers, are being introduced for zirconia additive manufacturing [19–23].

The photopolymerization method shows relatively higher precision, faster manufacturing speed, and higher surface roughness among these techniques. Therefore, it is more likely to be used in manufacturing dental prostheses [24]. The DLP uses a projector, modified to produce a specific wavelength, and a digital micromirror device to cure the liquid resin in layer units. Although the precision is lower than the SLA method, using a laser scanner, DLP shows excellent accuracy in small prints, such as crowns, and is advantageous due to its high manufacturing speed and relatively inexpensive equipment cost [7,25].

In dentistry, strength and precision are enormous challenges when manufacturing zirconia prostheses by DLP additive manufacturing. The process requires a zirconia photopolymer suspension, generally including zirconia powder, photosensitive resin, photoinitiator, and dispersant. For zirconia to have high strength, a high zirconia volume fraction is required, creating difficulty in obtaining an appropriate degree of dispersion and viscosity [26]. After mixing, it is crucial to have a well-dispersed suspension with relatively low viscosity to ensure precise additive manufacturing [27,28]. A study by Jang et al. [29] on the production of zirconia using DLP additive manufacturing showed that a volume fraction of zirconia of 58 vol% was the maximum range possible to achieve homogeneous mixing. As the volume fraction increases, the 3-point bending strength increases. However, the study reported that the viscosity increased rapidly to 56 vol%.

Geometrical overgrowth was introduced as a factor affecting manufacturing accuracy using DLP additive manufacturing zirconia [29,30]. Geometrical overgrowth refers to light scattering during the photopolymerization of the zirconia suspension, resulting in the over-curing of the surrounding. Light scattering occurs due to the high refractive index, polycrystalline grain structure, and relatively large grain size of zirconia [29–31]. A previous study [29,32] reported the evaluation of curing depth and geometrical overgrowth of additive manufacturing according to the volume fraction of zirconia and showed that the geometrical overgrowth by light scattering increased as the volume fraction increased, and a decreased curing time caused light scattering. A reduction in curing time could adversely affect cure depth. The photopolymerization method in ceramic additive manufacturing showed that overgrowth increased as the exposure time and area increased [30,33]. Based on the previous report, once the slurry and printing parameters were optimized, the layer lines played a minor role in the strength [34], which eventually highlights the importance of printing overgrowth optimization. Literature shows few laboratory studies on the geometrical overgrowth of additive manufacturing of zirconia. Mitteramskogler et al. [30,35] added a UV absorber to the suspension composition to suppress overgrowth due to light scattering during ceramic additive manufacturing. The UV absorber absorbs UV rays and converts them into thermal energy [36]. It can also control the penetration depth of UV rays and prevent proper dispersion and over-curing [37–39]. While the UV absorber is mainly studied as a polymer stabilizer by adding it to the photopolymer suspension, further studies on its effectiveness as a light scattering inhibitor and controlling overgrowth in zirconia additive manufacturing are needed.

Additive manufacturing using commercially available photopolymer resin for 3D printing not only controls but also compensates for the curing contraction of the resin and has shown clinically acceptable precision for manufacturing crowns and other prostheses [27]. It is necessary to compensate for the shrinkage due to sintering and evaluate and control geometrical overgrowth to obtain stability and precision in dental prostheses during the additive manufacturing of zirconia prostheses. Therefore, this study aimed to assess the degree of geometrical overgrowth based on the UV absorber of DLP additive manufacturing zirconia and also looks to increase the accuracy of additively manufactured zirconia prostheses in dental practice. Moreover, we aimed to assess the influence of UV absorbers on the dimensional accuracy of zirconia shaping fabricated by DLP additive manufacturing.

2. Materials and Methods

2.1. Zirconia Suspension Preparation

Zirconia photopolymer suspension was prepared based on the acrylate series commercialized with zirconia powder (TZ-3Y; Tosoh, Tokyo, Japan), and Table 1 shows the mechanochemical properties of each material [29]. The other additive agents included a photoinitiator (Irgacure 819; Ciba Specialty Chemicals, Basel, Switzerland) and dispersant (BYK-180; BYK Inc., Wesel, Germany) added based on the previous study [29]. The volume fraction of the zirconia was calculated to be 54 vol% to prepare a suspension. Trimethoxysilane (MTMS; Sigma-Aldrich Inc., Saint Louis, MO, USA) was added for silane treatment, and hydroxyphenyl-triazine (Tinuvin-477; BASF, Ludwigshafen, Germany) was added as a UV absorber (Orange 3, Sigma-Aldrich, Saint Louis, MO, USA) [40] Table 2 shows the composition of zirconia suspension for each group. The UV absorber volume fraction was divided into four groups, with added 0, 0.005, 0.05, and 0.1 vol%, respectively. A planetary centrifugal mixer (ARV-310; Thinky Corp., Tokyo, Japan) was used for homogenous mixing.

Table 1. Materials and their properties.

	Materials	Density (g/mL)	Refractive Index (325 nm)	Viscosity (25 °C, mPas)
Ceramic	Zirconia	6.05	2.15	-
Photopolymer	HDDA *	1.02	1.45	9
	IBA **	0.98	1.47	8
	PBPGDA ***	1	1.44	15
Additive agents	Photoinitiator	1.19–1.21	-	-
	Dispersant	1.075	-	-
Silane coupling agent	MTMS ****	0.955	1.371	-
UV absorber	Hydroxyphenyl-triazine	1.08	-	-

* 1,6-Hexanediol diacrylate, ** Isobomyl aceylate, *** Propoxylated neopentyl glycol diacrylatemtms, and **** Methyltrimethoxysilane.

Table 2. Composition of zirconia suspension.

Group	Zirconia	Photopolymer	Silane Coupling Agent	Dispersant	UV Absorber	Total (Vol%)
o	54	27.62	6.28	12.1	0	100
A5		27.60			0.005	100
A50		27.43			0.05	100
A100		27.24			0.1	100

2.2. Fabrication of Specimen via Additive Manufacturing

Figure 1 shows the specimen for this study was designed in nine hollow structures as a 2 mm square of 20 mm × 20 mm × 1 mm using CAD software (SolidWorks 2016, Dassault Systems Corp., Vélizy-Villacoublay, France) was then converted to a standard tessellation language (STL) file for additive manufacturing. Zirconia specimens were prepared using DLP equipment (Octave Light R1, Octave Light Ltd., Shatin, Hongkong) (wavelength 365–405 nm), setting the thickness of each layer to 50 μm and the exposure time to 4 s ($n = 5$) [31].

2.3. Geometrical Overgrowth Evaluation

Geometrical overgrowth occurs by curing a specimen into a wider area than the existing area exposed to the light source resulting from the light scattering effect of zirconia particles [31]. This experiment measured the degree of reduction of the nine hollow square structures. The overgrowth observed in the experiment was compared to the STL file standard in length and area. Figure 2 shows the length was measured using a stereomicroscope (EGVM-452M; EG Tech, Gwangyang, South Korea) by one person at a specific position throughout the study (n = 5).

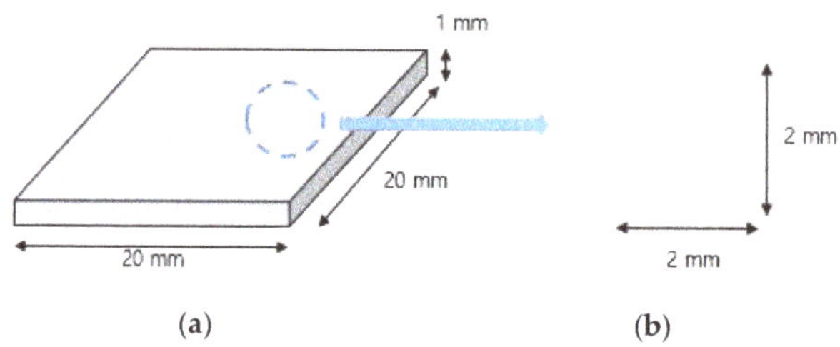

Figure 1. (**a**) Schematization of the STL file of the specimen. (**b**) The square hole (2 × 2 mm) for geometrical overgrowth measurement.

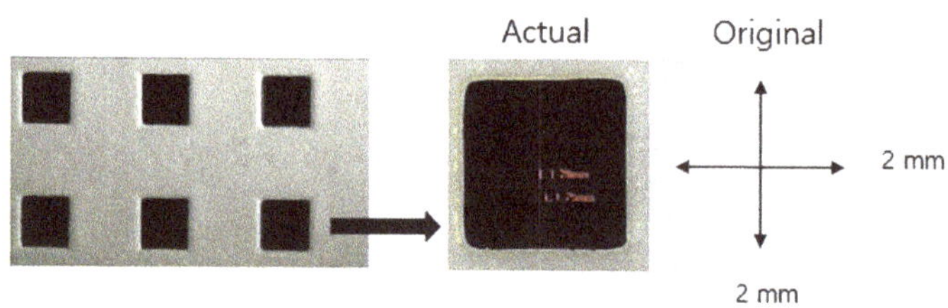

Figure 2. Measurement of linear deviation using a stereoscopic microscope.

Figure 3 shows the area measured using image analysis software (ImageJ software, National Institutes of Health, Bethesda, MD, USA). Geometrical overgrowth was quantified by calculating the difference between the actual measured and the designed size (2 mm).

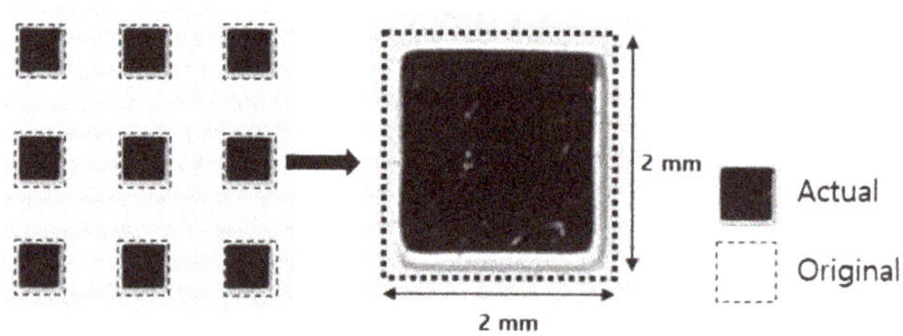

Figure 3. Measurement of area deviation using image analysis software.

2.4. Cure Depth

Cure depth was measured to evaluate the effect of the UV absorber during lamination. Under the same conditions as the DLP 3D printer, a distance of 20 cm, a light of 100 mW/cm^2, and a light exposure time of 4 s were maintained. Figure 4 shows the final cured thickness (n = 3) based on the volume fraction of the UV absorber, which was measured using digital vernier calipers (n = 3).

2.5. Microstructural Analysis

The conditions for degreasing and sintering were set based on the results of previous experiments [39]. The microstructure and surface of the specimens were observed before and after sintering, and the defects were observed using a field emission scanning electron microscope (FE-SEM; JEOL) (ISO-13356) based on the UV absorber volume fraction [38].

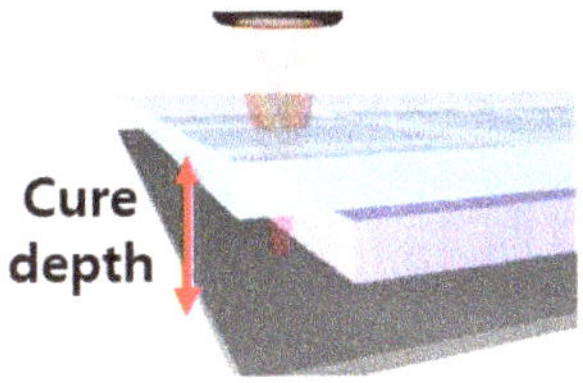

Figure 4. Schematic representation of curing depth of zirconia suspension by UV exposure.

2.6. Statistical Analysis

All statistical analyses were performed using SPSS 21.0 (SPSS Inc.; Chicago, IL, USA). All results were tested for significance at the $p < 0.0083$ level. As normality was not satisfied in the Shapiro–Wilk test, the Kruskal–Wallis, a nonparametric test, was performed. The post hoc test was performed using the Bonferroni correction method.

3. Result and Discussion

3.1. Geometrical Overgrowth Evaluation

It is essential to evaluate the strength and precision of additive manufacturing of zirconia dental prostheses using DLP. Geometric overgrowth, caused by over-curing resulting from the light scattering effect, is a factor that can affect precision. Light scattering is affected by the volume fraction of zirconia, the difference in refractive index between the zirconia powder and the medium, and the curing time [28–30]. Jang et al. [29] reported that zirconia has a refractive index of 2.1, which is 20–27% higher than silica and alumina. Thus, the degree of light scattering is large, and the cure depth is limited. In this study, a UV absorber was added to the 54 vol% zirconia photopolymer suspension to evaluate the degree of overgrowth. The cure depth was measured to ensure sufficient curing for additive manufacturing.

Figures 5 and 6 and Table 3 show the degree of overgrowth by subtracting the length and area of each specimen measured manufactured based on the UV absorber ratio by the designed value. In all the groups, the void structure decreased and showed an overgrowth pattern. As the UV absorber ratio increased, the degree of overgrowth tended to decline. In terms of length, the average degree of overgrowth was 40 μm in the A100 group and 251 μm in the O group. Regarding the ratio, the length showed an overgrowth rate of 2.1% in the A100 group and 12.5% in the O group, and the area showed an overgrowth rate of 8.5% in the A100 group and 12.25% in the O group. Significant differences were observed in the length and areas of the O group and the A50 and A100 groups without a UV absorber ($p < 0.0083$). No significant difference was observed based on the UV absorber ratio.

The results highlight that the group prepared by adding 0.05% and 0.1% UV absorber to the zirconia suspension (A50 group, A100 group) showed a significantly greater degree of overgrowth in length and area than the group without the addition of UV absorber (Group O). Our previous study [29] measured geometrical overgrowth according to the volume fraction of zirconia, showing an overgrowth rate of 33.52% at 54 vol%, higher than the results obtained in the current study. Despite the differences in the composition of the UV absorber and suspension, a geometrical overgrowth can be considered a factor that influences the larger exposure area and longer cure time than the specimen in this study. Geometrical overgrowth can be affected by the presence or absence of a UV absorber, as shown in this study, in addition to factors such as curing time, light quantity, exposure area, and difference in suspension composition [7,24,29,30].

The results show that the geometrical overgrowth was the lowest in the A100 group, containing 0.1% UV absorber, with overgrowth rates of 2.1% and 8.5% in length and area, respectively. When the overgrowth rate of the length was lower, the overgrowth was not constant by region, and the periphery of the rectangle tended to be rounded. Thus, we assumed that there was a difference in the area when the length was measured at a certain intermediate position. Additional studies are needed on the degree of overgrowth and the direction and pattern of overgrowth according to the specimen shape for precise additive manufacturing of zirconia.

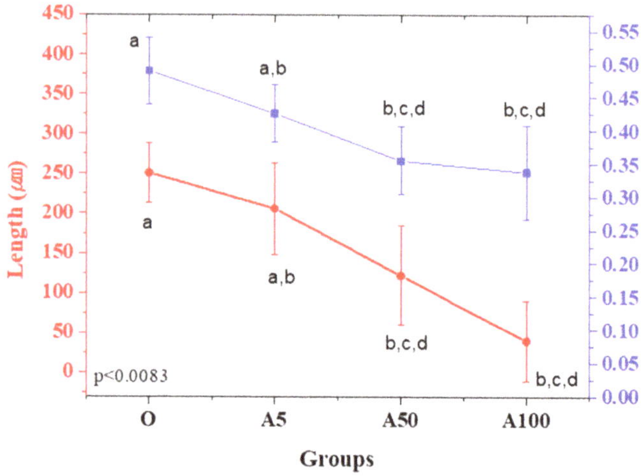

Figure 5. Comparison of length (μm) and area (mm^2) of the prepared sample. Different letters indicate significant differences and when the same letters included there is no significant differences. ($p < 0.0083$).

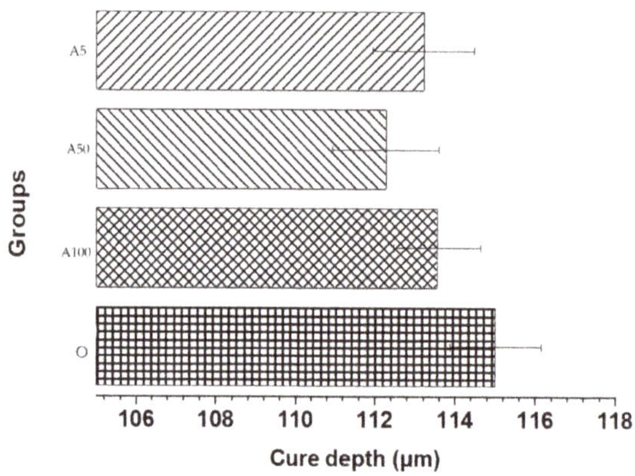

Figure 6. The bar graph represents the cure depth (μm) of the prepared sample.

3.2. Cure Depth

Zirconia prostheses are generally manufactured by setting the layer thickness between 25 and 100 μm during DLP additive processing [17]. The pre-sintered body is additively manufactured according to the fixed layer thickness. If the curing depth is lower than the layer thickness, delamination may occur between layers, causing micro-defects and deterioration of properties [7]. Therefore, attempts to suppress light scattering by adding a UV absorber or adjusting the curing time should consider whether the cure depth is sufficient beyond the set layer thickness.

In this study, the thickness of the laminated layer was set to 100 μm and measured to check whether adding a UV absorber affects layering. Figure 6 and Table 4 show there was no significant difference in the cure depth based on the UV absorber. All groups showed a cure depth of 100 μm or more. The cure depth was obtained at a uniform intensity of light and distance and measured using digital vernier calipers (n = 3). Surface defects and microstructures were analyzed using a field emission scanning electron microscope (FE-SEM; JEOL). The specimens were sintered at 1450 °C to observe the surface after sintering.

3.3. Microstructural Analysis

Figure 7 shows the cross-sectional microstructure of the zirconia specimen before and after sintering between the final selected A100 group specimen and the control group O group specimen. It confirmed that a 100 μm layer and slight microcracks were visible before sintering in both groups (Figure 7a). It is judged that the microcracks are generated during the cleaning process after printing is completed, suggesting caution is needed even during the post-treatment process. In Figure 7b, the cross-section of each specimen after sintering was observed. After sintering, it was confirmed that each printing layer confirmed before sintering had healing and simultaneous defects. It is judged that defects occurred due to the characteristic of zirconia shrinking during heat treatment. Compared to the control group, the lower cure depths of A100 did not provide adhesion to cause perfect healing between each layer. Figure 7c shows the high magnification of each sintered specimen. The grain sizes of 0.46 (±0.06) and 0.58 (±0.083) nm of the A100 group and O group can be confirmed, and also, it is confirmed that group O shows a difference of about 100 nm. However, there was no significant difference in each group. It is due to similar grain growth and powder characteristics since the experiment was performed under the same degreasing and sintering conditions.

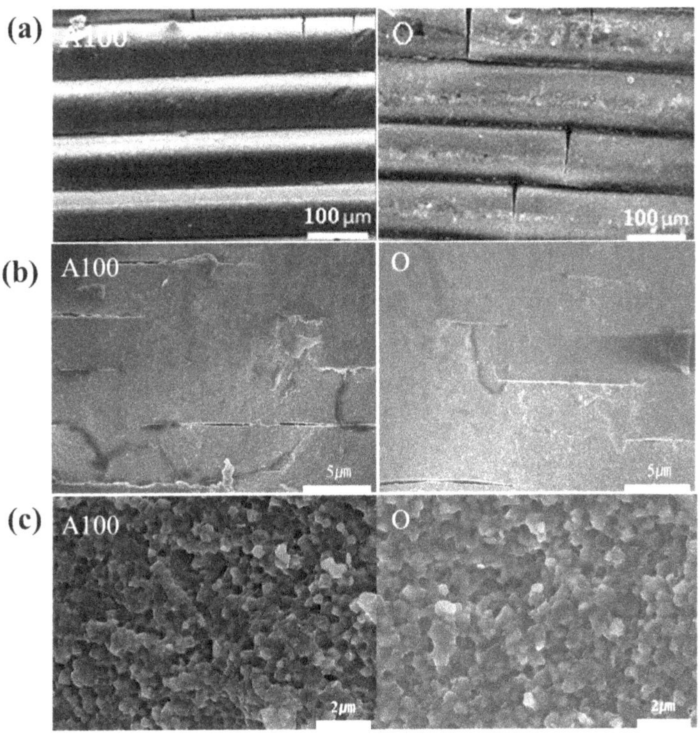

Figure 7. FE-SEM surface image at magnifications of (**a**) green bodies, (**b**) sintered specimens and (**c**) sintered specimens of high magnification.

Table 3. The mean and standard deviation of length and area.

Group	Length (μm)	Area (mm^2)
A5	20 ± 657.4	0.426 ± 0.043
A50	122 ± 62.3	0.355 ± 0.051
A100	40 ± 50.4	0.338 ± 0.071
O	251 ± 37.1	0.490 ± 0.050

Table 4. The mean and standard deviation of cure depth.

Group	Cure Depth (μm)
A5	113.55 ± 1.09
A50	112.27 ± 1.33
A100	113.21 ± 1.27
O	115.01 ± 1.14

4. Conclusions

This study investigated the degree of geometrical overgrowth when a UV absorber was added to the additive manufacturing of zirconia using DLP. When evaluating the overgrowth rate, the length measurement ranged from 2.1% to 12.5%, and the area measurement ranged from 8.5% to 12.25%. Results showed a significant difference in the overgrowth between the groups without the UV absorber (Group O) and the group with 0.05% and 0.1% (A50 group and A100 group). The degree of overgrowth was the smallest in the A100 group and the highest in the O group. Length measurement revealed an average value of 40 μm in the A100 group and 251 μm in the O group. The average area was 0.490 mm^2 in group O. There was no significant difference in the cure depth in the groups with and without the UV absorber. Cure depths of 100 μm or more were observed in all groups. No surface defects or laminated zirconia layers were observed in the specimen after sintering in all groups. Thus, the geometrical overgrowth of pre-sintered bodies decreased when the UV absorber was added to the DLP-based additive manufacturing of zirconia, suggesting the possibility of increased accuracy of additive manufacturing of zirconia. Thus, sintering shrinkage compensation, reflecting geometrical overgrowth, is necessary to increase the accuracy in the additive manufacturing of zirconia.

Author Contributions: Conceptualization and writing—original draft preparation, S.-W.P., J.-H.K. and K.S.; writing—review and editing, S.-W.P., J.-H.K., K.S., C.P. and K.-D.Y.; investigation and data curation, H.-A.K., M.-J.J. and H.-P.L.; methodology, H.-A.K. and S.A.H.T.; visualization, S.A.H.T. and C.P.; supervision, S.-W.P. and K.-D.Y.; funding acquisition, J.-H.K. and S.-W.P. All authors have read and agreed to the published version of the manuscript.

Funding: This work was supported by a National Research Foundation of Korea Grant funded by the Korean government (MSIP) (Grant no. 2022R1A2C20096331140982119420101 and 2022R1A6A3A010870061140982119420101).

Institutional Review Board Statement: Not applicable.

Informed Consent Statement: Not applicable.

Data Availability Statement: All the data have been illustrated in the manuscript.

Conflicts of Interest: The authors declare no conflict of interest.

References

1. Beuer, F.; Schweiger, J.; Edelhoff, D. Digital dentistry: An overview of recent developments for CAD/CAM generated restorations. *Br. Dent. J.* **2008**, *204*, 505–511. [CrossRef] [PubMed]
2. Zhang, X.; Wu, X.; Shi, J. Additive manufacturing of zirconia ceramics: A state-of-the-art review. *J. Mater. Res. Technol.* **2020**, *9*, 9029–9048. [CrossRef]
3. Lian, Q.; Wu, X.; Li, D.; He, X.; Meng, J.; Liu, X.; Jin, Z. Accurate printing of a zirconia molar crown bridge using three-part auxiliary supports and ceramic mask projection stereolithography. *Ceram. Int.* **2019**, *45*, 18814–18822. [CrossRef]

4. Osman, R.B.; van der Veen, A.J.; Huiberts, D.; Wismeijer, D.; Alharbi, N. 3D-printing zirconia implants; a dream or a reality? An in-vitro study evaluating the dimensional accuracy, surface topography and mechanical properties of printed zirconia implant and discs. *J. Mech. Behav. Biomed. Mater.* **2017**, *75*, 521–528. [CrossRef]
5. Xing, B.; Cao, C.; Zhao, W.; Shen, M.; Wang, C.; Zhao, Z. Dense 8 mol% yttria-stabilized zirconia electrolyte by DLP stereolithography. *J. Eur. Ceram. Soc.* **2020**, *40*, 1418–1423. [CrossRef]
6. Takahashi, M.; Kirihara, S. Stereolithographic Additive Manufacturing of Zirconia Electrodes with Dendritic Patterns for Aluminum Smelting. *Appl. Sci.* **2021**, *11*, 8168. [CrossRef]
7. Zhang, J.; Wei, L.; Meng, X.; Yu, F.; Yang, N.; Liu, S. Digital light processing-stereolithography three-dimensional printing of yttria-stabilized zirconia. *Ceram. Int.* **2020**, *46*, 8745–8753. [CrossRef]
8. He, R.; Liu, W.; Wu, Z.; An, D.; Huang, M.; Wu, H.; Jiang, Q.; Ji, X.; Wu, S.; Xie, Z. Fabrication of complex-shaped zirconia ceramic parts via a DLP-stereolithography-based 3D printing method. *Ceram. Int.* **2018**, *44*, 3412–3416. [CrossRef]
9. Wang, L.; Liu, X.; Wang, G.; Tang, W.; Li, S.; Duan, W.; Dou, R. Partially stabilized zirconia moulds fabricated by stereolithographic additive manufacturing via digital light processing. *Mater. Sci. Eng. A* **2020**, *770*, 138537. [CrossRef]
10. Wang, J.-C.; Dommati, H. Fabrication of zirconia ceramic parts by using solvent-based slurry stereolithography and sintering. *Int. J. Adv. Manuf. Technol.* **2018**, *98*, 1537–1546. [CrossRef]
11. Ohkuma, K.; Kameda, T.; Terada, K. Five-axis laser milling system that realizes more accurate zirconia CAD/CAM crowns by direct milling from fully sintered blocks. *Dent. Mater. J.* **2019**, *38*, 52–60. [CrossRef] [PubMed]
12. Manicone, P.F.; Iommetti, P.R.; Raffaelli, L. An overview of zirconia ceramics: Basic properties and clinical applications. *J. Dent.* **2007**, *35*, 819–826. [CrossRef] [PubMed]
13. Lebon, N.; Tapie, L.; Duret, F.; Attal, J.-P. Understanding dental CAD/CAM for restorations-dental milling machines from a mechanical engineering viewpoint. Part A: Chairside milling machines. *Int. J. Comput. Dent.* **2016**, *19*, 45–62. [PubMed]
14. Denry, I.; Kelly, J.R. State of the art of zirconia for dental applications. *Dent. Mater.* **2008**, *24*, 299–307. [CrossRef]
15. Preis, V.; Behr, M.; Hahnel, S.; Handel, G.; Rosentritt, M. In vitro failure and fracture resistance of veneered and full-contour zirconia restorations. *J. Dent.* **2012**, *40*, 921–928. [CrossRef]
16. Park, J.-M.; Ahn, J.-S.; Cha, H.-S.; Lee, J.-H. Wear resistance of 3D printing resin material opposing zirconia and metal antagonists. *Materials* **2018**, *11*, 1043. [CrossRef]
17. Dawood, A.; Marti, B.M.; Sauret-Jackson, V.; Darwood, A. 3D printing in dentistry. *Br. Dent. J.* **2015**, *219*, 521–529. [CrossRef]
18. Stansbury, J.W.; Idacavage, M.J. 3D printing with polymers: Challenges among expanding options and opportunities. *Dent. Mater.* **2016**, *32*, 54–64. [CrossRef]
19. Deckers, J.; Vleugels, J.; Kruth, J.-P. Additive manufacturing of ceramics: A review. *J. Ceram. Sci. Technol.* **2014**, *5*, 245–260.
20. Shahzad, K.; Deckers, J.; Zhang, Z.; Kruth, J.-P.; Vleugels, J. Additive manufacturing of zirconia parts by indirect selective laser sintering. *J. Eur. Ceram. Soc.* **2014**, *34*, 81–89. [CrossRef]
21. Ferrage, L.; Bertrand, G.; Lenormand, P. Dense yttria-stabilized zirconia obtained by direct selective laser sintering. *Addit. Manuf.* **2018**, *21*, 472–478. [CrossRef]
22. Song, X.; Chen, Y.; Lee, T.W.; Wu, S.; Cheng, L. Ceramic fabrication using Mask-Image-Projection-based Stereolithography integrated with tape-casting. *J. Manuf. Process.* **2015**, *20*, 456–464. [CrossRef]
23. Li, W.; Ghazanfari, A.; McMillen, D.; Leu, M.C.; Hilmas, G.E.; Watts, J. Characterization of zirconia specimens fabricated by ceramic on-demand extrusion. *Ceram. Int.* **2018**, *44*, 12245–12252. [CrossRef]
24. Halloran, J.W. Ceramic stereolithography: Additive manufacturing for ceramics by photopolymerization. *Annu. Rev. Mater. Res* **2016**, *46*, 19–40. [CrossRef]
25. Anssari Moin, D.; Hassan, B.; Wismeijer, D. A novel approach for custom three-dimensional printing of a zirconia root analogue implant by digital light processing. *Clin. Oral Implant. Res.* **2017**, *28*, 668–670. [CrossRef]
26. Tay, B.; Evans, J.; Edirisinghe, M. Solid freeform fabrication of ceramics. *Int. Mater. Rev.* **2003**, *48*, 341–370. [CrossRef]
27. Quan, H.; Zhang, T.; Xu, H.; Luo, S.; Nie, J.; Zhu, X. Photo-curing 3D printing technique and its challenges. *Bioact. Mater.* **2020**, *5*, 110–115. [CrossRef]
28. Kang, J.-H.; Sakthiabirami, K.; Jang, K.-J.; Jang, J.-G.; Oh, G.-J.; Park, C.; Fisher, J.G.; Park, S.-W. Mechanical and biological evaluation of lattice structured hydroxyapatite scaffolds produced via stereolithography additive manufacturing. *Mater. Des.* **2022**, *214*, 110372. [CrossRef]
29. Jang, K.-J.; Kang, J.-H.; Fisher, J.G.; Park, S.-W. Effect of the volume fraction of zirconia suspensions on the microstructure and physical properties of products produced by additive manufacturing. *Dent. Mater.* **2019**, *35*, e97–e106. [CrossRef]
30. Mitteramskogler, G.; Gmeiner, R.; Felzmann, R.; Gruber, S.; Hofstetter, C.; Stampfl, J.; Ebert, J.; Wachter, W.; Laubersheimer, J. Light curing strategies for lithography-based additive manufacturing of customized ceramics. *Addit. Manuf.* **2014**, *1*, 110–118. [CrossRef]
31. Jang, K.-J.; Kang, J.-H.; Sakthiabirami, K.; Lim, H.-P.; Yun, K.-D.; Yim, E.-K.; Oh, G.-J.; Yang, H.-S.; Lee, K.-K.; Park, S.-W. Evaluation of cure depth and geometrical overgrowth depending on zirconia volume fraction using digital light processing. *J. Nanosci. Nanotechnol.* **2019**, *19*, 2154–2157. [CrossRef]
32. Kang, J.-H.; Jang, K.-J.; Sakthiabirami, K.; Oh, G.-J.; Jang, J.-G.; Park, C.; Lim, H.-P.; Yun, K.-D.; Park, S.-W. Mechanical properties and optical evaluation of scaffolds produced from 45S5 bioactive glass suspensions via stereolithography. *Ceram. Int.* **2020**, *46*, 2481–2488. [CrossRef]

33. De Blas Romero, A.; Pfaffinger, M.; Mitteramskogler, G.; Schwentenwein, M.; Jellinek, C.; Homa, J.; Díaz Lantada, A.; Stampfl, J. Lithography-based additive manufacture of ceramic biodevices with design-controlled surface topographies. *Int. J. Adv. Manuf. Technol.* **2017**, *88*, 1547–1555. [CrossRef]
34. Coppola, B.; Schmitt, J.; Lacondemine, T.; Tardivat, C.; Montanaro, L.; Palmero, P. Digital light processing stereolithography of zirconia ceramics: Slurry elaboration and orientation-reliant mechanical properties. *J. Eur. Ceram. Soc.* **2022**, *42*, 2974–2982. [CrossRef]
35. Cai, P.; Guo, L.; Wang, H.; Li, J.; Li, J.; Qiu, Y.; Zhang, Q.; Lue, Q. Effects of slurry mixing methods and solid loading on 3D printed silica glass parts based on DLP stereolithography. *Ceram. Int.* **2020**, *46*, 16833–16841. [CrossRef]
36. Mau, R.; Nazir, J.; Seitz, H. Dimensional accuracy of 3D printing of PEGDA parts using Digital Light Processing technology. *Trans. Addit. Manuf. Meets Med.* **2019**, *1*, S03P11.
37. Li, Y.; Mao, Q.; Li, X.; Yin, J.; Wang, Y.; Fu, J.; Huang, Y. High-fidelity and high-efficiency additive manufacturing using tunable pre-curing digital light processing. *Addit. Manuf.* **2019**, *30*, 100889. [CrossRef]
38. Sakornwimon, N.; Leevailoj, C. Clinical marginal fit of zirconia crowns and patients' preferences for impression techniques using intraoral digital scanner versus polyvinyl siloxane material. *J. Prosthet. Dent.* **2017**, *118*, 386–391. [CrossRef]
39. Jang, J.-G.; Kang, J.-H.; Joe, K.-B.; Sakthiabirami, K.; Jang, K.-J.; Jun, M.-J.; Oh, G.-J.; Park, C.; Park, S.-W. Evaluation of Physical Properties of Zirconia Suspension with Added Silane Coupling Agent for Additive Manufacturing Processes. *Materials* **2022**, *15*, 1337. [CrossRef]
40. Gmeiner, R.; Mitteramskogler, G.; Stampfl, J.; Boccaccini, A.R. Stereolithographic ceramic manufacturing of high strength bioactive glass. *Int. J. Appl. Ceram. Technol.* **2015**, *12*, 38–45. [CrossRef]

materials MDPI

Article
Fracture Resistance of CAD/CAM Implant-Supported 3Y-TZP-Zirconia Cantilevers: An In Vitro Study

Mariana Novais, António Sérgio Silva *, Joana Mendes, Pedro Barreiros, Carlos Aroso and José Manuel Mendes

UNIPRO–Oral Pathology and Rehabilitation Research Unit, University Institute of Health Sciences (IUCS), CESPU, 4585-116 Gandra, Portugal
* Correspondence: asergio.silva@iucs.cespu.pt; Tel.: +351-224157100

Abstract: (1) Introduction: Implant-supported fixed complete dentures are mostly composed of cantilevers. The purpose of this work was to evaluate the fracture resistance of zirconia (Prettau®, second generation, or Ice Zirkon Translucent, first generation) with cantilever lengths of 6 and 10 mm, and zirconia's fracture resistance in relation to an average bite force of 250 N. (2) Materials and methods: Forty structures were created in CAD/CAM and divided into four groups: group A (6 mm cantilever in IZT), group B (10 mm cantilever in IZT), group C (6 mm cantilever in Pz), and group D (10 mm cantilever in pz). The study consisted of a traditional "load-to-failure" test. (3) Results: A statistically significant result was found for the effect of cantilever length, $t(38) = 16.23$ ($p < 0.001$), with this having a large effect size, $d = 4.68$. The 6 mm cantilever length ($M = 442.30$, $sd = 47.49$) was associated with a higher mean force at break than the 10 mm length ($M = 215.18$, $sd = 40.74$). No significant effect was found for the type of zirconia: $t(38) = 0.31$ ($p = 0.757$), and $d = 0.10$. (4) Conclusions: All the components with cantilever lengths of 6 mm broke under forces higher than 250 N. Cantilevers larger than 10 mm should be avoided.

Keywords: cantilever length; fracture; bite force; CAD/CAM; implant-supported prosthesis; zirconia

Citation: Novais, M.; Silva, A.S.; Mendes, J.; Barreiros, P.; Aroso, C.; Mendes, J.M. Fracture Resistance of CAD/CAM Implant-Supported 3Y-TZP-Zirconia Cantilevers: An In Vitro Study. *Materials* **2022**, *15*, 6638. https://doi.org/10.3390/ma15196638

Academic Editors: Sangwon Park and John G. Fisher

Received: 31 August 2022
Accepted: 20 September 2022
Published: 24 September 2022

1. Introduction

The global elderly population will reach about 400 million in 2050, [1] which will increase partial or total edentulism. In the case of complete dentures, their stability and retention can be obtained through dental implants allowing fixed rehabilitation, or through the oral mucosa, in which dental adhesives, in the case of removable rehabilitation, play a fundamental role [2,3]. Fixed prostheses on implants are prostheses that have stability and retention, supported by dental implants, and they have been used for more than three decades. Osseointegrated dental implants have revolutionized prosthetic treatments and have become a fundamental alternative in the quality of life of populations [2–5]. However, implant-supported fixed rehabilitation can cause problems due to the anatomical and morphological conditions of the patient, which condition the selection and distribution of implants [2]. Computer-aided design and computer-aided manufacturing (CAD–CAM) can be used in different production techniques, namely, the subtractive milling and additive manufacturing. Regarding these two techniques, Valenti et al. showed that there is no significant difference between them in terms of the hardness, roughness, marginal discrepancy, fracture load, trueness, or internal fit. Furthermore, the additive manufacturing does not allow continuous mastication forces for a long period of time, being mostly used in provisional crowns and fixed partial dentures [6].

This is a determining factor in the design of prosthetic structures. Therefore, implant-supported fixed complete dentures, commonly known as hybrid prostheses, are mostly composed of cantilevers [7], that is, a multiple retainer with one or more unsupported free ends [8], thus allowing the prosthesis to extend to at least the first molar [2]. The use of cantilevers in fixed prostheses on dental implants is beneficial in places with unfavorable anatomical characteristics, namely, reduced alveolar ridges [9]; however, the placement of implants can be compromised due to their proximity to certain structures, such as the maxillary sinuses, the roots of adjacent teeth, and the inferior alveolar nerve [8]. However, long cantilevers should be avoided due to their biomechanical properties, which can lead to prosthesis fracture [5,7] as well as loss of bone around the implants [5]. As such, it is recommended that mandibular cantilevers do not exceed 20 mm, and ideally, they should be less than 15 mm in length [8]. Approximately half of patients rehabilitated with fixed prostheses containing cantilevers have long-term complications [2]. These can be due to high occlusal load, but also due to poorly distributed occlusal forces, which can lead to the loss or fracture of implants used to retain dental prostheses, which happens most often at the beginning of the arm of the cantilever. In order to avoid this type of fracture, shortening the mesiodistal, buccolingual/palatal length of the cantilever is recommended [2]. Thus, for a rehabilitation with implants to be successful, it is necessary that implant-supported prostheses are resistant to fracture [10].

Aesthetics plays a very important role in our society; as such, oral rehabilitation must take into account aesthetic requirements by using materials that meet the expectations of patients [11,12]. However, this should not compromise the strength, clinical success, and longevity of dental prostheses. The stress caused during their use can be transferred to the implants, to the bone, or to the supporting structures, which highlights the importance of the choice of material [13,14]. Zirconia differs from other materials due to the phenomenon called transformation hardening, with three pure forms of zirconia having been identified, depending on the temperature at which it is found [11].

Due to their aesthetic properties, there has been a substantial increase in the use of zirconia and ceramics in prosthetic rehabilitation [9,11,12]. This increase is due not only to the aesthetic component, but also due to these materials' high fracture resistance and other mechanical properties [11], their excellent biocompatibility, and their physical properties [9,11,15]. These materials can be used in the elaboration of crowns, posts, fixed partial dentures, abutments, and structures on implants [9]. Currently, yttrium-stabilized tetragonal zirconia (Y-TZP) is the ceramic material that has the highest fracture resistance [16], and it is the ceramic material most used as a rehabilitation material in high stress areas due to its high strength [15]. It should be noted that today, the elaboration of zirconia structures is more optimized, accurate, and reliable, due to the development of computerized CAD–CAM systems [12,17–20]. They may be a viable alternative for full-arch implant-supported rehabilitations, being able to achieve high 5-year survival rates with minimal prosthetic complication rates since they were not subject to chipping and wear, even if work needs to be performed to improve their esthetics [21,22]. However, their medium- to long-term clinical outcomes cannot yet be evaluated [23–25].

Finite element analysis studies [26] showed that when force was applied over the cantilevers, the highest stress was concentrated in the distal posterior screw-access openings (SAOs), which is the zone with the minimum cross-sectional connector area (CSCA) because of the screw opening passing through it. This means that distal SAOs are the zones that are supposed to fracture when excessive strength is applied over the cantilever. This type of prosthesis distal extension fracture is directly related to the cantilever [27]. The other type of fractures are fractures, that occur between the distal SAOs and not involving them, are hypothesized to be more likely concerned with other causes, such as tension in the screw-retained structure, material defects, or excessive thinness of the zirconia framework [26,27].

An incidence of 5.6% prosthetic failure rate is described in most of the articles, but clinically we can observe a higher percentage that is not described [28].

Thus, this work has as its main objective to evaluate the fracture resistance of different cantilever lengths (CL) (6 and 10 mm). It also attempts to determine which of the zirconia forms (Prettau® or Ice Zirkon Translucent) of lengths of 6 mm and 10 mm are more resistant to fracture, and to evaluate the fracture resistance of zirconia under an average bite force of

250 N. The null hypothesis tested is that the frameworks with a cantilever length of 6 mm have the same fracture force of the frameworks with a cantilever length of 10 mm. The limits of this study are those understood for an in vitro study and the application of force was performed only in one direction, excluding the application of oblique forces.

2. Materials and Methods

2.1. Materials

All materials used in this study were selected based on their importance and usefulness in dentistry, as well as their stability under normal conditions of use and storage. All materials and chemicals were used in accordance with manufacturers' standards.

The materials used in this study were two different types of zirconia that belong to the same brand. One of the materials used was Ice Zirkon Translucent (IZT) (Zirkonzahn®, Gais, South Tyrol, Italy), and the other was Prettau® Zirconia (PZ) (Zirkonzahn®, Gais, South Tyrol, Italy), as described in Table 1.

Table 1. Zirconia characteristics.

Zirconia	Flexural Strength	Vickers Hardness (HV10)
Ice Zirkon Translucent	1200–1400 MPa	1250 HV10
Prettau® Zirconia	1000–1200 MPa	1250 HV10

The interfaces used, as well as the screws, were from IPD® (Mataró, Barcelona, Spain). These interfaces were selected because they are compatible with internal hexagon implants.

2.2. Methods

A standard laboratory protocol was established and applied at the Laboratório de Investigação em Reabilitação Oral e Prostodontia, UNIPRO—Oral Pathology and Rehabilitation Research Unit, University Institute of Health Sciences (IUCS), CESPU, Gandra, Portugal, to test all selected samples.

2.2.1. Preparation of the Sample

In this study, 40 CAD/CAM frameworks were prepared and divided into four groups: group A (10 frameworks with a 6 mm cantilever in Ice Zirkon Translucent), group B (10 frameworks with a 10 mm cantilever in Ice Zirkon Translucent), group C (10 frameworks with a 6 mm cantilever in Prettau® Zirconia), and group D (10 frameworks with a 10 mm cantilever in Prettau® Zirconia) (Figure 1). All the structures were implant-supported by two internal hexagon analogs with a 4.1 platform, and fixed on a titanium base, with there being a distance of 15 mm from their centers. They were all manufactured 4 mm high and 3 mm wide by IPD® (Mataró, Barcelona, Spain). A titanium base was prepared so that it could be adapted to the support table to fix the testing machine, Instron®, Electropuls E10000 Linear-Torsion (Norwood, MA, USA), on which the two internal hexagon analogs with a 4.1 platform were coupled together.

The framework has a rectangle milled shape with linear polished edges.

2.2.2. Elaboration of Zirconia Structures

The structures were digitally executed using a CAD/CAM system (Zirkonzahn® Gais, South Tyrol, Italy) (Figure 2a,b). Two body scans (IPD®, Mataró, Barcelona, Spain) were placed on the base analogues and they were read in the Scanner S600 (Zirkonzahn® Gais, South Tyrol, Italy) in order to be able to use a digital library and thus ensure that the structure would be well adapted to the interfaces (Figure 2c). Interfaces with a height of 3.5 mm were used.

After the structures were digitally executed, they were milled using the M1 machine from Zirkonzahn® (Gais, South Tyrol, Italy). After this, they were removed from the zirconia blocks using a turbine and diamond drill, removing only the zirconia supports. The frameworks were sintered in a Zirkonofen 600 EV oven (Zirkonzahn®, Gais, South

Tyrol, Italy). The IZT structures were placed in the oven at a maximum temperature of 1400 °C (gradual rise of 8 °C/min, 2 h at maximum temperature, and cooling of 8 °C/min until room temperature was reached), under which their dimensions contracted by 20%. In turn, the PZ structures were placed in the oven at a maximum temperature of 1600 °C (gradual rise of 6 °C/min, 2 h at maximum temperature, and cooling of 6 °C/min until room temperature was reached), under which they contracted in size by 19.95%.

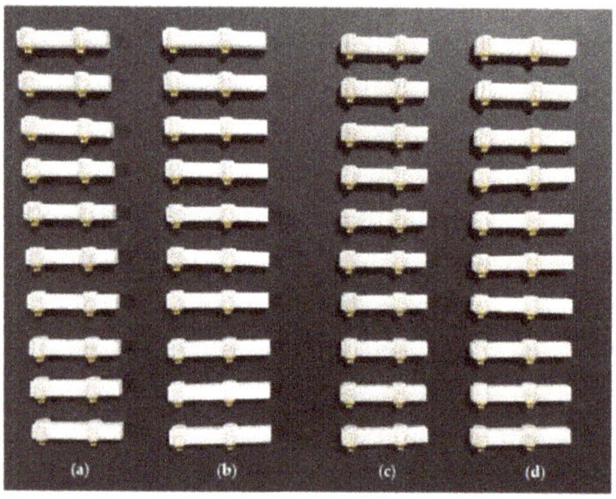

Figure 1. Study structures: (**a**) 6 mm Ice Zirkon Translucent cantilever; (**b**) 10 mm Ice Zirkon Translucent cantilever; (**c**) 6 mm Prettau® Zirconia cantilever; (**d**) 10 mm Prettau® Zirconia cantilever.

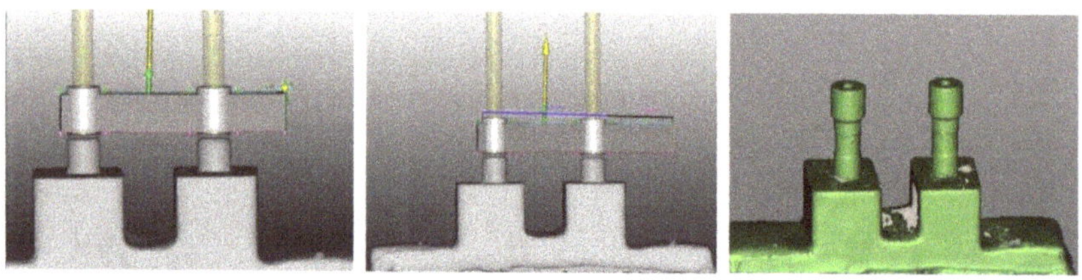

(**a**) (**b**) (**c**)

Figure 2. CAD/CAM elaboration of zirconia frameworks: (**a**) 6 mm cantilever; (**b**) 10 mm cantilever; (**c**) body scans.

In order to complete the test structures, the interfaces were screwed to the base (previously prepared) at 35 N using a torque wrench. The structures were cemented to the interfaces with Maxcem EliteTM resin cement (Kerr®, Kloten, Switzerland), following the manufacturer's instructions. Teflon tape was placed in the access channel to the interface screws to prevent the entry of cement. After polymerization, all excesses were removed.

2.2.3. Compression Test to Measure the Fracture Resistance Strength of Different Cantilevers

The titanium base, with the implant analogs made for the study, was attached to the Instron® fixation support table, as shown, allowing its connection to the Instron® testing machine, the Electropuls E10000 Linear-Torsion.

The Instron® Electropuls E10000 LT is a dynamic and fatigue testing machine with linear dynamic capacity of ±10 KN, linear static capacity of ±7 KN, linear stroke of 60 mm, torque capacity of ±100 Nm, torsion stroke of ±135°, and daylight 877 mm opening, which allows both static and dynamic axial and torsional tests, in accordance with ISO 7500-1:2018. It has a calibration accredited force according to ISO 7500-1 and ASTM E4 standard of up to 5 meganewtons. The fixture support table was attached to the machine to adapt the simulation structures, so that all the models were adjusted and produced equal compression. In this way, the test structures were parallel to the transport table.

The test structures (bars with cantilevers and interfaces) were screwed to the titanium base, as shown in Figure 3a, at 35 N using a torque wrench. The study consisted of a traditional load-to-failure test in which a static load was used, 2 mm from the end of the cantilever arm (Figure 3b), which progressively increased by 1 mm/min towards the structure until the fracture occurred (Figure 3c,d), as described in Alshiddi et al.'s study [9].

Figure 3. Fracture resistance tests. (**a**) Titanium-based bolted structure; (**b**) start of the static load, 2 mm from the end of the cantilever arm; (**c**) fracture of the 6 mm cantilever; (**d**) fracture of the 10 mm cantilever.

The number of samples that was used for each zirconia type was 10 replicates (10 cantilevered frameworks) with 6 mm and 10 mm cantilever lengths. In this way, the fracture resistance force of each type of structure was measured 10 times, and the values were recorded in newtons (N).

Test results were transferred to WaveMatrix® version 2.0 Dynamic Test Software (Instron®, Norwood, MA, USA). This software allows users to define and run tests and acquire data for a wide variety of dynamic and quasi-static applications. Then, all the values and data were transferred to Microsoft Office Excel®, version 16.0 (Redmond, WA, USA) where the statistical analysis of the obtained data was performed.

2.3. Statistical Analysis

Data were analyzed with R, version 4.1.2 (R Core Team, R Foundation for Statistical Computing, Vienna, Austria). Initial data inspection included normality assessment of force at break (standard) measured in newtons (N), using the Shapiro–Wilk test, which is appropriate for sample sizes lower than 50. Levene's test was used to assess the homogeneity of variance. Considering the results of the tests, parametric tests were implemented. Descriptive statistics are presented as means (M) and standard deviations (SD). Boxplots were also used to plot the distribution of force at break, with points added that corresponded to the sampling observations. Observation overlap was avoided by including a jitter function. Independent samples *t*-tests were used to compare force at break according to the type of zirconia (Translucent and Prettau), cantilever length (6 mm and 10 mm), and the type of zirconia stratified by each cantilever length. Cohen's d was calculated for effect size, with 0.2 considered a small effect size, 0.5 a medium effect size, and 0.8 a large effect size.

The interaction between the type of zirconia and cantilever length was assessed using factorial ANOVA. Partial eta squared ($\eta_p{}^2$) was used to assess effect size, with 0.01 considered a small effect size, 0.06 a medium effect size, and 0.14 a large effect size

One-sample *t*-tests were used to compare the mean forces at break under the limit of 250 N.

Statistical significance was considered for $p < 0.05$. A higher threshold (marginal significance, $p < 0.10$) was considered when the sample size was 10 or lower.

3. Results

A total of 40 component samples were enrolled in a 2 × 2 design that considered the type of zirconia (Translucent and Prettau) and the cantilever length (6 mm and 10 mm). The resistance to fracture was assessed with the force at break (standard) measured in newtons (N). Table 2 and Figure 4 show results for the independent effects of the type of zirconia and cantilever length on force at break. A statistically significant result was found for the effect of cantilever length, t(38) = 16.23 ($p < 0.001$), with this having a large effect size, d = 4.68. Lengths of 6 mm (M = 442.30, sd = 47.49) were associated with a higher mean force at break when compared with lengths of 10 mm (M = 215.18, sd = 40.74). No significant effect was found for the zirconia type: t(38) = 0.31 ($p = 0.757$), d = 0.10.

Table 2. Force at break (standard) (N) according to type of zirconia and cantilever length.

	Type of Zirconia			Cantilever Length		
	Translucent (n = 20)	Prettau (n = 20)	*t*-Test	6 mm (n = 20)	10 mm (n = 20)	*t*-Test
Force at Break (Standard) (N)	322.62 (137.09)	334.86 (110.41)	t(38) = 0.31 ($p = 0.757$) d = 0.10	442.30 (47.49)	215.18 (40.74)	t(38) = 16.23 ($p < 0.001$) d = 4.68

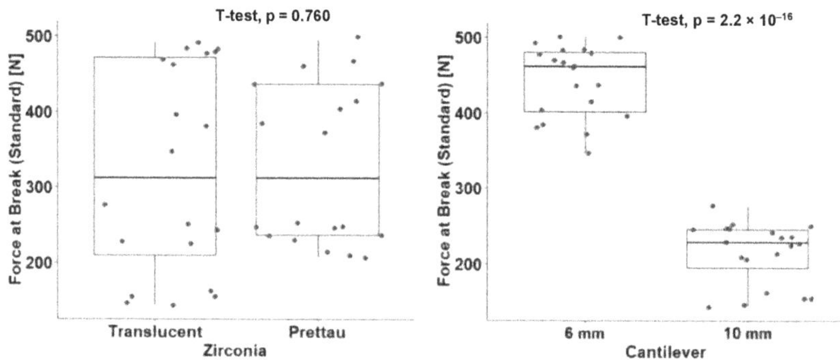

Figure 4. Force at break (standard) (N) according to type of zirconia and cantilever length.

Table 3 and Figure 5 show results for the independent effect of the type of zirconia stratified by cantilever length (6 mm and 10 mm). For the 10 mm length, a marginal effect was detected, t(18) = −2.01 ($p = 0.070$), with this having a large effect size, d = 0.90. The Prettau Zirconia components showed a higher mean force at break (M = 232.18, sd = 16.83) than the Ice Zirkon Translucent components (M = 198.18, sd = 50.79). No marginal or significant effect was detected for the 6 mm cantilever length.

Table 3. Force at break (standard) (N) according to type of zirconia stratified by cantilever length (6 mm and 10 mm).

	Cantilever Length = 6 mm (n = 20)			Cantilever Length = 10 mm (n = 20)		
	Translucent Zirconia (n = 10)	Prettau Zirconia (n = 10)	*t*-Test	Translucent Zirconia (n = 10)	Prettau Zirconia (n = 10)	*t*-Test
Force at Break (Standard) (N)	447.05 (51.84)	437.54 (44.97)	t(18) = 0.44 ($p = 0.667$) d = 0.20	198.18 (50.79)	232.18 (16.83)	t(18) = −2.01 ($p = 0.070$) ‡ d = 0.90

‡ $p < 0.10$.

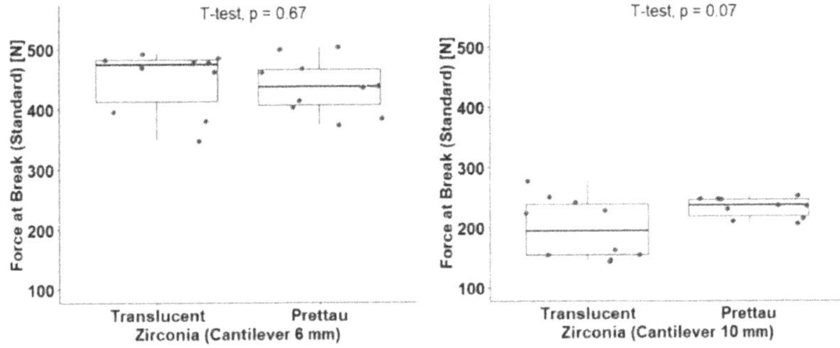

Figure 5. Force at break (standard) (N) according to type of zirconia stratified by cantilever length (6 mm and 10 mm).

Next, the interaction between the type of zirconia and cantilever length was studied by conducting a factorial ANOVA (Table 4, Figure 6). The initial effects of the type of zirconia and cantilever length were assessed. Cantilever length showed a significant effect, $F_{(1,36)} = 272.46$ ($p < 0.001$), a large effect size, $\eta_p^2 = 0.88$, and a higher mean force at break for 6 mm, as previously shown. No significant effect was found for the type of zirconia: $F_{(1,36)} = 0.79$ ($p = 0.379$), and $\eta_p^2 = 0.02$. The interaction between the type of zirconia and cantilever length was also not statistically significant: $F_{(1,36)} = 2.50$ ($p = 0.273$), and $\eta_p^2 = 0.07$, meaning that the cantilever length effect of 6 mm was independent of the type of zirconia.

Table 4. Factorial ANOVA for the effect of type of zirconia and cantilever length on force at break (standard) (N).

		n	M	SD	Factorial ANOVA
Zirconia	Cantilever				
Translucent	6 mm	10	447.05	51.84	Zirconia effect $F_{(1,36)} = 0.79$ ($p = 0.379$), $\eta_p^2 = 0.02$
	10 mm	10	198.18	50.79	
	Total	20	322.62	137.09	
Prettau	6 mm	10	437.54	44.97	Cantilever effect $F_{(1,36)} = 272.46$ ($p < 0.001$), $\eta_p^2 = 0.88$
	10 mm	10	232.18	16.83	
	Total	20	334.86	110.41	
Total	6 mm	20	442.30	47.49	Zirconia x Cantilever effect $F_{(1,36)} = 2.50$ ($p = 0.273$), $\eta_p^2 = 0.07$
	10 mm	20	215.18	40.74	
	Total	40	328.74	123.02	

Finally, force at break was compared to the limit of 250 N under the null hypothesis of H_0: $\mu \leq 250$ N vs. the alternative of H_1: $\mu > 250$ N. Comparisons were made that considered all samples as a whole and by group (Table 5). Unilateral one-sample *t*-tests showed statistically significant differences ($p < 0.001$) for all samples of a 6 mm cantilever length, regardless of the zirconia type, indicating that the required force to break the components is higher than 250 N for components with a cantilever length of 6 mm. All components with a 6 mm cantilever length broke under forces higher than 250 N. On the contrary, for the cantilever length of 10 mm, the null hypothesis was not rejected ($p > 0.999$) for both Ice Zirkon Translucent, with just two (20%) components of which broke under forces higher than 250 N, and Prettau Zirconia, with just 1 (10%) component of which broke after applying forces higher than 250 N.

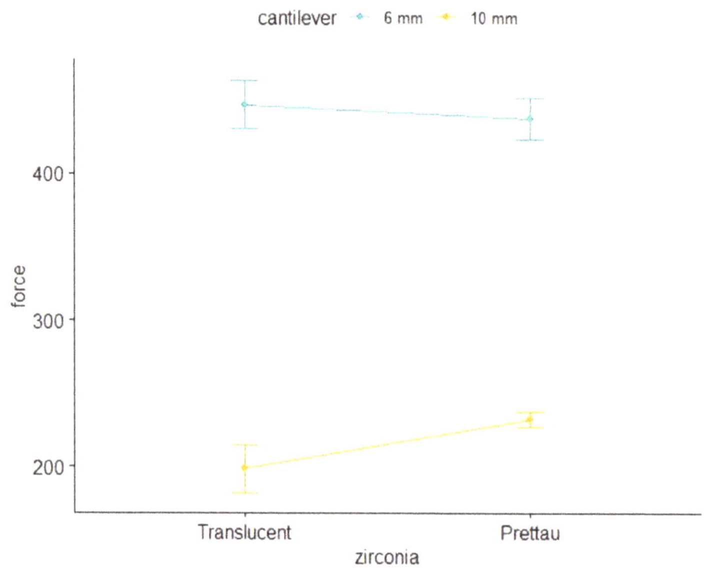

Figure 6. Force at break (standard) (N) interaction between type of zirconia and cantilever length.

Table 5. Unilateral one-sample *t*-tests for assessing limits of force at break (standard) (N).

		Tests for Mean > 250 N				Proportion of Components > 250 N
		n	M	SD	*t*-Tests	n (%)
Zirconia	Cantilever					
Translucent	6 mm	10	447.05	51.84	t(9) = 12.02 ($p < 0.001$)	10 (100%)
	10 mm	10	198.18	50.79	t(9) = −3.23 ($p > 0.999$)	2 (20%)
	Total	20	322.62	137.09	t(19) = 2.37 ($p = 0.015$)	12 (60.0%)
Prettau	6 mm	10	437.54	44.97	t(9) = 13.19 ($p < 0.001$)	10 (100%)
	10 mm	10	232.18	16.83	t(9) = −3.35 ($p > 0.999$)	1 (10%)
	Total	20	334.86	110.41	t(19) = 3.44 ($p = 0.002$)	11 (55.0%)
Total	6 mm	20	442.30	47.49	t(19) = 18.11 ($p < 0.001$)	20 (100%)
	10 mm	20	215.18	40.74	t(19) = −3.82 ($p > 0.999$)	3 (15.0%)
	Total	40	328.74	123.02	t(39) = 4.05 ($p < 0.001$)	23 (57.5%)

Figure 7 shows that all the components included in the study needed to be submitted to forces higher than 250 N to break. As previously shown, all components with a cantilever length of 6 mm required a force higher than 250 N to break. In the whole sample, the minimum and maximum forces at break were 143.07 N and 501.58 N, respectively.

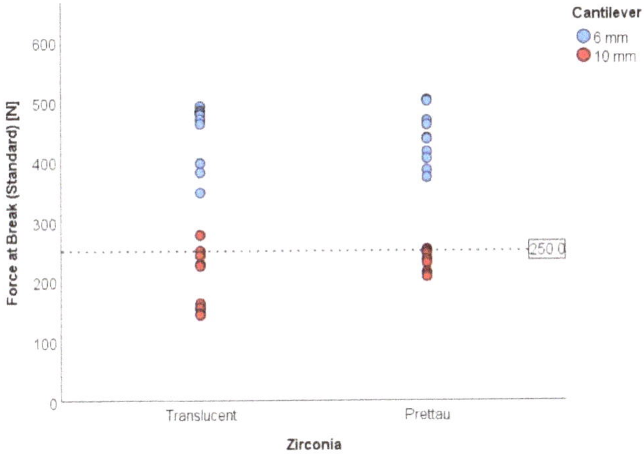

Figure 7. Force at break (standard) (N) for all components included in the study and the considered limits.

4. Discussion

For fixed implant rehabilitations, namely, rehabilitations with structures with cantilevers, the CL is a very important factor to take into account, as it influences the longevity of the rehabilitation. In this study, we tested the fracture resistance of different lengths of cantilevers, 6 and 10 mm, and also the resistance of two different types of zirconia, Ice Zirkon Translucent and Prettau® Zirconia. For this purpose, 40 frameworks were made and tested: 20 frameworks with a 6 mm cantilever, of which 10 were in Ice Zirkon Translucent and 10 were in Prettau® Zirconia, and 20 frameworks with a 10 mm cantilever, of which 10 were also in Ice Zirkon Translucent and in Prettau® zirconia.

Tirone et al. reported on the CL or the thickness of the zirconia around the screw access opening (SAO) to the fracture of the structure [29], and their results were in line with those found in our study, because the SAO was the zone where there was a fracture in the zirconia structures. In order to overcome this problem, Alshiddi et al. used a thickness increase of 0.5 mm around the distal SAO in order to reinforce it [9]. In our study, the value used around the SAO was the predefined value of the CAD/CAM program, namely, 0.5 mm, in order not to modify the values that are established for use in the elaboration of structures of implant-supported zirconia prostheses.

With regard to CL, the results obtained in this study corroborate what is described by several authors [2,9,29]. Bearing in mind that the ideal cantilever length should not exceed 9 mm, in this study, structures with 6 mm and 10 mm cantilevers were tested in order to test their resistance to fracture. A higher force was required to cause fracture in 6 mm structures (a mean force of 442.30 N) than in 10 mm structures (a mean force of 215.18 N). It was also found that when a force was applied to the cantilever, the greatest stress point was concentrated in the distal SAO, with there being a smaller cross-sectional connector area due to access to the SAO, these being the areas where fracture is expected.

Bite force (BF) is an important component of mastication as well as masticatory function [30]. The measurement of BF allows us to determine the average BF values of a given population, thus helping us choose the most suitable and resistant materials during prosthetic rehabilitation. Based on Takaki et al.'s study, in which they reported a mean BF of 285.01 N for women and 253.99 N for men [31], and Levartovsky et al.'s study, in which they reported an average BF of between 258.5 N and 175.8 N [32], we used an average BF reference value of 250 N. Taking into account this average BF value and analyzing the results obtained, we can see that all structures with a 6 mm cantilever required a force greater than the average reference value for fracture to occur, unlike structures with a 10 mm cantilever, of which only three structures resisted forces greater than 250 N (one in

PT and two in IZT). However, Van Vuuren et al. reported an average BF of 430.4 N [33], and, taking this value into account, we can see that only structures with a 6 mm cantilever can resist this average BF, given that the mean fracture values found in this study were 437.54 N in PT and 447.05 N in IZT. Taking into account these mean BF values and based on the results of this study, we can say that in terms of longevity of prosthetic rehabilitations, structures with 6 mm cantilevers have a higher fracture resistance behavior than those found in cantilevers 10 mm long. Regarding the type of zirconia used (Ice Zirkon Translucent and Prettau® Zirconia), they performed similarly in terms of fracture resistance, with there being no statistically significant differences between them.

Y-TZP, a widely developed ceramic rehabilitation material [34], is considered the most resistant and robust ceramic material available [35]. Clinically, stratified crowns after prolonged use have shown fissures or fractures, leading to rehabilitation failure [12,36,37]. This problem has been solved with the introduction and development of monolithic zirconia, which is manufactured in CAD/CAM. The first generation of 3Y-TZP was developed to be the strongest, accepting loads of up to 1200 MPa [24]. In this type of rehabilitation, all the ceramic layering is removed, thus preventing one of the reasons for its failure [34,36,37].

In this study, the differences in the average strength of materials were not considered significant. IZT presented an average fracture resistance of 322.62 N, while PZ presented an average fracture resistance of 334.86 N. However, even though the difference in the strength of the different materials was minimal, the fracture strength of the 10 mm cantilever PZ obtained better results than that of the IZT, the former requiring an average force of 232.18 N compared to the latter's 198.18 N. Meanwhile, the 6 mm cantilever PZ required an average force of 437.54, and the IZT required an average force of 447.05 N.

The framework shape was not evaluated in this study, but some studies suggest that geometry influences the material behavior and fracture, e.g., a conventional shape (linear, e.g., our framework design); convex shape (1.0 mm curve in the direction of the occlusal surface); concave shape (1.0 mm curve in the direction of the gingival surface). The different results, for example, in Tsumita et al.'s study of the mechanical strength in different framework shapes, showed that the fracture load for the convex shape was the highest; however, critical cracks in the veneer porcelain were seen in the convex shape, but not in the other two shapes [38]. These cracks occurred from the lower margin of the pontic framework towards the gingival surface of the medial and distal connectors of the pontic framework. Such failure is not clinically acceptable. Because of the geometry of the convex shape, it was difficult for the frameworks located on the gingival side of the connector where stress concentrates; thus, the veneer porcelain received the tensile stress directly, and cracks were initiated at a low load value. In addition, the final fracture load was high, because the shape of frameworks resembled a reverse catenary, and thus received the loading stress as compressive stress. The fracture load for the concave shape was significantly higher than that for the conventional shape; the veneer porcelain cracking load for the concave shape was significantly greater than that for the convex shape. In terms of bridge engineering, the concave shape resembled a catenary; however, by arranging a framework without a pontic–connector interface where stress concentrates in an area of maximum principal stress, the load could be evenly dispersed throughout the lower margin of the frame.

Conventionally cast noble (gold/silver–palladium) or not noble (Co–Cr) metal alloys are the most traditionally employed materials for full-arch implant-supported rehabilitations, reporting high clinical performances with optimal clinical implant and prosthetic survival rates in the long term. However, various alternative materials are available today, such as titanium, zirconia, and several polymers including carbon-fiber frameworks, providing corrosion resistance, biocompatibility, and great mechanical characteristics, with satisfactory clinical outcomes [28].

Further comparative clinical studies, possibly randomized clinical trials with a longer follow-up time, are needed in order to validate the use of new materials and define their specific clinical indications.

5. Conclusions

In the present study, it was found that there was a big difference in the fracture resistance of the cantilevers used, and, taking into account the average BF, fracture strength was much higher in the 6 mm cantilevers compared to the 10 mm cantilevers.

Regarding the type of zirconia, IZT or PT, we found that there were no statistically significant differences regarding their strength in the two cantilever lengths (6 and 10 mm).

Taking into account the average BF of 250 N, in the 6 mm cantilever, the IZT and PT did not show significant differences in fracture resistance, both presenting an average fracture force greater than 250 N. On the other hand, in the 10 mm cantilevers, the IZT and the PT obtained inferior results in relation to this mean BF value in terms of fracture resistance, with both presenting a lower mean fracture force.

Author Contributions: Conceptualization, M.N. and J.M.M.; methodology, A.S.S. and C.A.; software, P.B.; validation, J.M.M. and A.S.S.; formal analysis, J.M.; investigation, M.N.; resources, J.M.M.; data curation, A.S.S.; writing—original draft preparation, M.N.; writing—review and editing, J.M.; visualization, A.S.S. and C.A.; supervision, J.M.M.; project administration, A.S.S.; funding acquisition, P.B. All authors have read and agreed to the published version of the manuscript.

Funding: This research received no external funding.

Institutional Review Board Statement: Not applicable.

Informed Consent Statement: Not applicable.

Data Availability Statement: The data that support the findings of this study are available from the corresponding author upon request.

Acknowledgments: We are grateful for the availability and cooperation of the Department of Dental Sciences, the Cooperative for Polytechnic and University Education (Cooperativa de Ensino Superior Politécnico e Universitário—CESPU).

Conflicts of Interest: The authors declare no conflict of interest.

References

1. Petersen, P.E.; Kandelman, D.; Arpin, S.; Ogawa, H. Global oral health of older people–call for public health action. *Community Dent Health* **2010**, *27*, 257–268. [PubMed]
2. Minoretti, R.; Triaca, A.; Saulacic, N. Unconventional implants for distal cantilever fixed full-arch prostheses: A long-term evaluation of four cases. *Int. J. Periodontics Restor. Dent.* **2012**, *32*, e59–e67.
3. Mendes, J.; Mendes, J.M.; Barreiros, P.; Aroso, C.; Silva, A.S. Retention Capacity of Original Denture Adhesives and White Brands for Conventional Complete Dentures: An In Vitro Study. *Polymers* **2022**, *14*, 1749. [CrossRef]
4. Takaba, M.; Tanaka, S.; Ishiura, Y.; Baba, K. Implant-supported fixed dental prostheses with CAD/CAM-fabricated porcelain crown and zirconia-based framework. *J. Prosthodont.* **2013**, *22*, 402–407. [CrossRef] [PubMed]
5. Gonçalves, F.C.P.; Amaral, M.; Borges, A.L.S.; Gonçalves, L.F.M.; Paes-Junior, T.J.A. Fracture load of complete-arch implant-supported prostheses reinforced with nylon-silica mesh: An in vitro study. *J. Prosthet. Dent.* **2018**, *119*, 606–610. [CrossRef] [PubMed]
6. Valenti, C.; Isabella, F.M.; Masciotti, F.; Marinucci, L.; Xhimitiku, I.; Cianetti, S.; Pagano, S. Mechanical properties of 3D-printed prosthetic materials compared with milled and conventional processing: A systematic review and meta-analysis of in vitro studies [published online ahead of print, 2022 Aug 5]. *J. Prosthet. Dent.* **2022**, *S0022-3913*, 00415-2. [CrossRef]
7. Shackleton, J.L.; Carr, L.; Slabbert, J.C.; Becker, P.J. Survival of fixed implant-supported prostheses related to cantilever lengths. *J. Prosthet. Dent.* **1994**, *71*, 23–26. [CrossRef]
8. Chong, K.K.; Palamara, J.; Wong, R.H.; Judge, R.B. Fracture force of cantilevered zirconia frameworks: An in vitro study. *J. Prosthet. Dent.* **2014**, *112*, 849–856. [CrossRef]
9. Alshiddi, I.F.; Habib, S.R.; Zafar, M.S.; Bajunaid, S.; Labban, N.; Alsarhan, M. Fracture Load of CAD/CAM Fabricated Cantilever Implant-Supported Zirconia Framework: An In Vitro Study. *Molecules* **2021**, *13*, 2259. [CrossRef]
10. Honda, J.; Komine, F.; Kusaba, K.; Kitani, J.; Matsushima, K.; Matsumura, H. Fracture loads of screw-retained implant-supported zirconia prostheses after thermal and mechanical stress. *J. Prosthodont. Res.* **2020**, *3*, 313–318. [CrossRef]
11. Alraheam, I.A.; Donovan, T.; Boushell, L.; Cook, R.; Ritter, A.V.; Sulaiman, T.A. Fracture load of two thicknesses of different zirconia types after fatiguing and thermocycling. *J. Prosthet. Dent.* **2020**, *123*, 635–640. [CrossRef] [PubMed]
12. Fischer, P.; Barbu, H.M.; Fischer, C.A.I.; Pantea, M.; Baciu, F.; Vranceanu, D.M.; Cotrut, C.M.; Spinu, T.C. Bending Fracture of Different Zirconia-Based Bioceramics for Dental Applications: A Comparative Study. *Materials* **2021**, *14*, 6887. [CrossRef]

13. Bacchi, A.; Consani, R.L.; Mesquita, M.F.; Dos Santos, M.B. Effect of framework material and vertical misfit on stress distribution in implant-supported partial prosthesis under load application: 3-D finite element analysis. *Acta Odontol. Scand.* **2013**, *71*, 1243–1249. [CrossRef]
14. Ferreira, M.B.; Barão, V.A.; Faverani, L.P.; Hipólito, A.C.; Assunção, W.G. The role of superstructure material on the stress distribution in mandibular full-arch implant-supported fixed dentures. A CT-based 3D-FEA. *Mater. Sci. Eng. C Mater. Biol. Appl.* **2014**, *35*, 92–99. [CrossRef]
15. Hafezeqoran, A.; Koodaryan, R.; Hemmati, Y.; Akbarzadeh, A. Effect of connector size and design on the fracture resistance of monolithic zirconia fixed dental prosthesis. *J. Dent. Res. Dent. Clin. Dent. Prospect.* **2020**, *14*, 218–222. [CrossRef]
16. Schriwer, C.; Skjold, A.; Gjerdet, N.R.; Øilo, M. Monolithic zirconia dental crowns. Internal fit, margin quality, fracture mode and load at fracture. *Dent. Mater.* **2017**, *33*, 1012–1020. [CrossRef] [PubMed]
17. Chougule, K.J.; Wadkar, A.P. An In vitro Comparative Evaluation of Flexural Strength of Monolithic Zirconia after Surface Alteration Utilising Two Different Techniques. *J. Clin. Diagn. Res.* **2017**, *11*, ZC20–ZC23. [CrossRef] [PubMed]
18. Juntavee, N.; Juntavee, A.; Phattharasophachai, T. Fracture toughness of different monolithic zirconia upon post-sintering processes. *J. Clin. Exp. Dent.* **2021**, *13*, e1006–e1014. [CrossRef]
19. Chang, Y.T.; Wu, Y.L.; Chen, H.S.; Tsai, M.H.; Chang, C.C.; Wu, A.Y. Comparing the Fracture Resistance and Modes of Failure in Different Types of CAD/CAM Zirconia Abutments with Internal Hexagonal Implants: An In Vitro Study. *Materials* **2022**, *15*, 2656. [CrossRef]
20. The R Development Core Team. A Language and Environment for Statistical Computing. 2021. Available online: http://softlibre.unizar.es/manuales/aplicaciones/r/fullrefman.pdf (accessed on 28 April 2022).
21. Bidra, A.S.; Tischler, M.; Patch, C. Survival of 2039 complete arch fixed implantsupported zirconia prostheses: A retrospective study. *J. Prosthet. Dent.* **2018**, *119*, 220–224. [CrossRef]
22. Tischler, M.; Patch, C.; Bidra, A.S. Rehabilitation of edentulous jaws with zirconia complete-arch fixed implant-supported prostheses: An up to 4-year retrospective clinical study. *J. Prosthet. Dent.* **2018**, *120*, 204–209. [CrossRef] [PubMed]
23. Poggio, C.E.; Ercoli, C.; Rispoli, L.; Maiorana, C.; Esposito, M. Metal-free materials for fixed prosthodontic restorations. *Cochrane Database Syst. Rev.* **2017**, *12*, CD009606. [CrossRef]
24. Bidra, A.S.; Rungruanganunt, P.; Gauthier, M. Clinical outcomes of full arch fixed implant-supported zirconia prostheses: A systematic review. *Eur. J. Oral Implantol.* **2017**, *10*, 35–45. [PubMed]
25. Abdulmajeed, A.A.; Lim, K.G.; Närhi, T.O.; Cooper, L.F. Complete-arch implant-supported monolithic zirconia fixed dental prostheses: A systematic review. *J. Prosthet. Dent.* **2016**, *115*, 672–677. [CrossRef] [PubMed]
26. Ozan, O.; Kurtulmus-Yilmaz, S. Biomechanical comparison of different implant inclinations and cantilever lengths in all-on-4 treatment concept by three-dimensional finite element analysis. *Int. J. Oral Maxillofac. Implants* **2018**, *33*, 64–71. [CrossRef]
27. Alshahrani, F.A.; Yilmaz, B.; Seidt, J.D.; McGlumphy, E.A.; Brantley, W.A. A load-to-fracture and strain analysis of monolithic zirconia cantilevered frameworks. *J. Prosthet. Dent.* **2017**, *118*, 752–758. [CrossRef]
28. Delucchi, F.; De Giovanni, E.; Pesce, P.; Bagnasco, F.; Pera, F.; Baldi, D.; Menini, M. Framework Materials for Full-Arch Implant-Supported Rehabilitations: A Systematic Review of Clinical Studies. *Materials* **2021**, *14*, 3251. [CrossRef]
29. Tirone, F.; Salzano, S.; Rolando, E.; Pozzatti, L.; Rodi, D. Framework Fracture of Zirconia Supported Full Arch Implant Rehabilitation: A Retrospective Evaluation of Cantilever Length and Distal Cross-Sectional Connection Area in 140 Patients Over an Up-To-7 Year Follow-Up Period. *J. Prosthodont.* **2022**, *31*, 121–129. [CrossRef]
30. Al-Omiri, M.K.; Sghaireen, M.G.; Alhijawi, M.M.; Alzoubi, I.A.; Lynch, C.D.; Lynch, E. Maximum bite force following unilateral implant-supported prosthetic treatment: Within-subject comparison to opposite dentate side. *J. Oral Rehabil.* **2014**, *41*, 624–629. [CrossRef]
31. Takaki, P.; Vieira, M.; Bommarito, S. Maximum bite force analysis in different age groups. *Int. Arch. Otorhinolaryngol.* **2014**, *18*, 272–276. [CrossRef]
32. Levartovsky, S.; Peleg, G.; Matalon, S.; Tsesis, I.; Rosen, E. Maximal Bite Force Measured via Digital Bite Force Transducer in Subjects with or without Dental Implants—A Pilot Study. *Appl. Sci.* **2022**, *12*, 1544. [CrossRef]
33. van Vuuren, L.J.; Broadbent, J.; Duncan, W.; Waddell, J. Maximum voluntary bite force, occlusal contact points and associated stresses on posterior teeth. *J. R. Soc. N. Z.* **2020**, *50*, 132–143. [CrossRef]
34. Schönhoff, L.M.; Lümkemann, N.; Buser, R.; Hampe, R.; Stawarczyk, B. Fatigue resistance of monolithic strength-gradient zirconia materials. *J. Mech. Behav. Biomed. Mater.* **2021**, *119*, 104504. [CrossRef]
35. Zhang, Y.; Lawn, B.R. Novel Zirconia Materials in Dentistry. *J. Dent. Res.* **2018**, *97*, 140–147. [CrossRef]
36. Tang, Z.; Zhao, X.; Wang, H.; Liu, B. Clinical evaluation of monolithic zirconia crowns for posterior teeth restorations. *Medicine* **2019**, *98*, e17385. [CrossRef]
37. Pjetursson, B.E.; Sailer, I.; Latyshev, A.; Rabel, K.; Kohal, R.J.; Karasan, D. A systematic review and meta-analysis evaluating the survival, the failure, and the complication rates of veneered and monolithic all-ceramic implant-supported single crowns. *Clin. Oral Implants Res.* **2021**, *32* (Suppl. S21), 254–288, Erratum in *Clin. Oral Implants Res.* **2021**, *32*, 1507. [CrossRef] [PubMed]
38. Tsumita, M.; Kokubo, Y.; Steyern, P.V.; Fukushima, S. Effect of Framework Shape on the Fracture Strength of Implant-Supported All-Ceramic Fixed Partial Dentures in the Molar Region. *J. Prosthodont.* **2008**, *17*, 274–285. [CrossRef] [PubMed]

materials
MDPI

Article

Effect of Non-Thermal Plasma Treatment of Contaminated Zirconia Surface on *Porphyromonas gingivalis* Adhesion and Osteoblast Viability

Seon-Ki Lee [1,†], Min-Kyung Ji [2,†], Yu-Jin Jo [3,†], Chan Park [3], Hoonsung Cho [4,*] and Hyun-Pil Lim [3,*]

1 Department of Prosthodontics, Daejeon Dental Hospital, Wonkwang University, Daejeon 35233, Korea; leesunki@gmail.com
2 Dental 4D Research Center, Chonnam National University, 33 Yongbong-ro, Buk-gu, Gwangju 61186, Korea; nada-juliet@hanmail.net
3 Department of Prosthodontics, School of Dentistry, Chonnam National University, 33 Yongbong-ro, Buk-gu, Gwangju 61186, Korea; nyota66@gmail.com (Y.-J.J.); upgradepc@hanmail.net (C.P.)
4 School of Materials Science & Engineering, Chonnam National University, 33 Yongbong-ro, Buk-gu, Gwangju 61186, Korea
* Correspondence: cho.hoonsung@jnu.ac.kr (H.C.); mcnihil@jnu.ac.kr (H.-P.L.); Tel.: +82-62-530-1717 (H.C.); +82-62-530-5577 (H.-P.L.)
† These authors contributed equally to this work.

Abstract: Plasma treatment on a zirconia surface prevents bacterial contamination and maintains osteoblast activity. To assess the degree of adhesion of *Porphyromonas gingivalis* on a zirconia surface after non-thermal plasma (NTP) treatment, specimens were treated with plasma for 60, 300, and 600 s, after which *P. gingivalis* was inoculated onto the surface and incubated for 48 h. To assess osteoblast activity after NTP treatment, osteoblasts (MC3T3-E1) were dispensed onto the specimens contaminated with *P. gingivalis* immediately after NTP for 60 and 120 s, followed by incubation for 48, 72, and 96 h. *P. gingivalis* was cultured after 60 s of NTP treatment of zirconia. The NTP and control groups showed no significant difference ($p = 0.91$), but adhesion was significantly increased following NTP treatment for 300 s or longer (300, 600 s groups) ($p < 0.05$). After NTP treatment of *P. gingivalis*-contaminated zirconia, osteoblast activity significantly increased at 72 and 96 h (I60 and I120 s group) in the groups treated with plasma ($p < 0.017$). Application of NTP to dental zirconia implants for 60 s not only inhibits the proliferation of *P. gingivalis*, which causes peri-implantitis but also increases osseointegration on zirconia surfaces contaminated with *P. gingivalis*.

Keywords: zirconia implant; non-thermal plasma; peri-implantitis; inhibition of biofilm formation; osteoblast viability

Citation: Lee, S.-K.; Ji, M.-K.; Jo, Y.-J.; Park, C.; Cho, H.; Lim, H.-P. Effect of Non-Thermal Plasma Treatment of Contaminated Zirconia Surface on *Porphyromonas gingivalis* Adhesion and Osteoblast Viability. *Materials* **2022**, *15*, 5348. https://doi.org/10.3390/ma15155348

Academic Editor: Guocheng Wang

Received: 22 June 2022
Accepted: 2 August 2022
Published: 3 August 2022

1. Introduction

Titanium and its alloys are widely used in dental implants [1]. However, because of adverse immunological reactions to titanium and the cosmetic limitations of metals, there is a growing need for new implant materials [2–4]. Zirconia can be used in patients with titanium allergy and shows it has advantages over titanium for use in oral areas where aesthetics require the use of a material with a color similar to that of natural teeth [5–7]. In addition, compared to titanium implants, zirconia has excellent biocompatibility and shows a lower adhesion of bacteria, thereby reducing the risk of peri-implantitis [8–12].

Periodontitis-related bacteria colonize exposed implants in the oral cavity within two weeks of implantation, and the entire range of subgingival flora develops within four weeks. Unlike natural teeth, implant prostheses do not possess periodontal ligaments, and the implant surface and alveolar bone are osseointegrated. Therefore, they are vulnerable to bacterial infection, leading to a high rate of alveolar bone loss in the presence of peri-implantitis, an inflammatory condition similar to periodontitis in which biofilms are formed on implant surfaces [13,14]. A biofilm consists of several species of bacterial flora, *Porphyromonas gingivalis, Aggregatibacter actinomycetemcomitans*, and *Fusobacterium nucleatum,* known as "red complexes", participate in late colonization [15]. Particularly, biofilms related to failed implants comprise gram-negative bacilli [16]. Among them, *P. gingivalis* is a gram-negative bacillus that influences plaque maturation and is mainly observed in well-differentiated subgingival plaques. Therefore, measuring *P. gingivalis* levels in advanced peri-implantitis can predict the prognosis of implants.

Atmospheric-pressure plasma can be used to treat peri-implantitis [17]. Non-thermal atmospheric pressure plasmas, also known as cold atmospheric plasmas or low-temperature atmospheric pressure plasmas, can be generated under atmospheric conditions and at low temperatures, facilitating their clinical application to the dental treatment area [18,19]. When plasma is applied to biomaterials, chemical changes lead to alterations in surface properties such as hydrophilicity, as the oxidized layer on the surface changes based on the chemical composition of the process gas, without changing the surface shape [20]. Yoo et al. reported changes in the shape of a colony of cultured bacteria on a plasma-applied titanium surface, with more than 80% of the cells killed [21], whereas Duske et al. [22] reported that plasma treatment increased the initial adhesion of osteoblasts and promoted bone fusion, by reducing the contact angle of the surface, which increased hydrophilicity.

Most studies have been performed in vitro by applying plasma to titanium [17]. Studies of plasma treatment of zirconia have focused on improving physical properties such as increasing the bonding strength of restoration [23,24]. Recently, changes in biological properties during the plasma treatment of zirconia were evaluated, including studies on cell activity [25], and improving osseointegration [26,27]. However, zirconia implants, peri-implantitis associated with zirconia implants, and inhibition of advanced peri-implantitis have not been widely examined.

Here, we investigated the effect of non-thermal plasma (NTP) treatment of zirconia surfaces on *P. gingivalis* adhesion and osteoblast activity in zirconia contaminated with *P. gingivalis* to determine the duration of plasma treatment that is clinically advantageous or both bacterial adhesion and osteoblast activity.

There were two null hypotheses. The first null hypothesis was that the adhesion of *P. gingivalis* would not differ after NTP treatment of zirconia. The second null hypothesis was that osteoblast adhesion would not differ in zirconia inoculated with *P. gingivalis*, even after NTP treatment of zirconia.

2. Materials and Methods

2.1. Materials

2.1.1. Samples

Fully sintered zirconia was prepared in disk-shaped specimens (diameter 15 mm, thickness 2.5 mm, Zirmon®, Kuwotech, Gwangju, Korea) and ground and polished with #800 SiC abrasive paper while injecting water in a polishing machine (LaboPol-5, Struers Co., Catcliffe Rotherham, UK) to give the specimens an even surface roughness. All specimens were washed for 20 min with acetone, alcohol, and distilled water using an ultrasonic cleaner. After drying at 25 °C, they were sterilized in an autoclave (HS-3460SD, Hanshin Medical Co., Incheon, Korea).

2.1.2. Artificial Saliva

To recreate the oral environment, artificial saliva was prepared by adding 1% mucin (M1778, Sigma-Aldrich, St. Louis, MO, USA) to the adherence buffer, as described by Li et al. [28].

2.1.3. Bacterial Culture

Porphyromonas gingivalis (KCOM 2804), an anaerobic, gram-negative anaerobic bacterium that causes peri-implantitis, was obtained from the Korean Collection for Oral Microbiology (Gwangju, Korea) and cultured. *P. gingivalis* strains in tryptic soy broth medium (BD Biosciences, Franklin Lakes, NJ, USA) were cultured at 37 °C in an anaerobic chamber (Forma Anaerobic System 1029; Thermo Fisher Scientific, Waltham, MA, USA). A single colony formed on the solid medium was inoculated into and cultured in a liquid medium to prepare bacteria at a concentration of 1.5×10^7 colony-forming units (CFU)/mL.

2.1.4. Osteoblast Culture

A mouse osteoblastic cell line (MC3T3-E1 Subclone 4, ATCC CRL2593, Rockville, MD, USA) was cultured at 37 °C in a CO_2 incubator at 5% CO_2 (Forma Series II 3111 Water Jacketed CO_2 Incubator, Thermo Fisher Scientific) inα-minimum essential Medium Gibco-BRL, Grand Island, NY, USA) containing 10% fetal bovine serum and 100 U/mL of penicillin. The culture medium was changed every three days, and the MC3T3-E1 cells were subcultured until enough cells were obtained for the experiment. Cells from generations 4–7 were used in the experiment.

2.2. *Methods*

2.2.1. NTP Treatment at Atmospheric Pressure

NTP at atmospheric pressure was applied to a zirconia specimen at a rate of 10 L/min and power of 300 W using a gaseous mixture containing Ar_2 (99.2%) and O_2 (0.8%) with an atmospheric pressure plasma generator (PGS-200, Expantech Co., Gyeonggi-do, Suwon, Korea). During plasma application, the length of the flame was maintained at 20 mm. The zirconia specimen was fixed in place with a circular jig and was rotated at 300 rpm. As the specimen rotated, the plasma alternated between left and right to ensure even application to the specimen. The reciprocation was set at 10 s per rotation and repeated automatically to match each instance of plasma application (Figure 1).

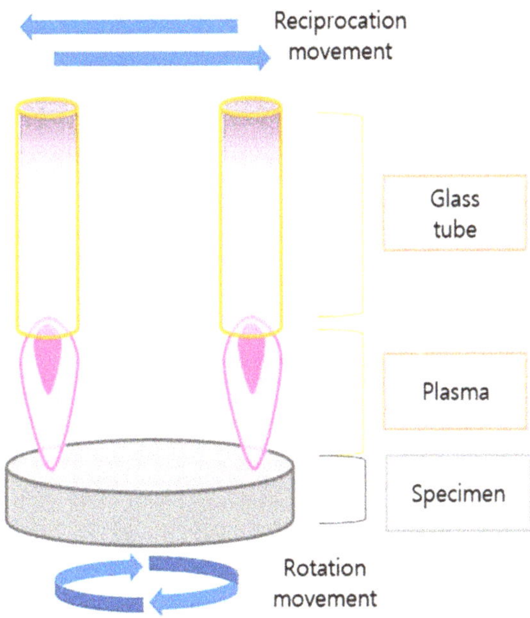

Figure 1. Even non-thermal plasma treatment of the specimen.

2.2.2. Assessment of Surface Characteristics

Energy-dispersive X-ray spectroscopy (EDX) (S-4700, Hitachi, Tokyo, Japan) and X-ray photoelectron spectroscopy (XPS) (VG Multilab 2000, Thermo Fisher Scientific) were performed to assess changes in the surface chemical composition of the specimens due to plasma treatment. To compare the changes in hydrophilicity, 4 μL of distilled water was dropped on the surface; after 10 s, the angle between the surface and water droplet was measured using a video contact angle meter (Phoenix 300, SEO, Kromtek, Selangor, DE, Malaysia). The average of the contact angles was calculated from the analysis of three specimens per group. Changes in the surface structure of the specimens, as well as bacterial and cell adhesion following the plasma treatment were observed using a scanning electron microscope (SEM) (S-4700, Hitachi) after coating the specimens with gold-palladium alloy using a Cressington sputter coater (108 auto, Cressington Scientific Instruments, Watford, UK).

2.2.3. Assessment of Bacterial Adhesion to Specimens after Plasma Treatment

All specimens were sterilized, and there were ten specimens in each experimental group. Bacterial adhesion to the specimen after plasma treatment was assessed in a crystal violet assay. Plasma-treated specimens were placed in a multi-well plate, and 1 mL of artificial saliva was dispensed into each well and stirred at room temperature for 2 h to coat the specimens. The artificial saliva was removed by aspiration and dried naturally for 15 min. Liquid medium (1.5×10^7 CFU/mL) was dispensed into each specimen and incubated for 48 h under anaerobic conditions. The specimens were carefully washed with phosphate-buffered saline (PBS) (Welgene, Daegu, Korea) after transferring to a new well plate, to measure the amount of *P. gingivalis* attached to the specimen surface. After staining with 0.3% crystal violet solution for 10 min, the solution was removed by aspiration and washed twice with PBS. After the specimens were naturally dried for 15 min, 400 μL of destaining solution (80% ethyl alcohol + 20% acetone) was dispensed onto the specimens and decolorized for 1 h. The destaining solution (200 μL) was transferred to a 96-well plate and the absorbance was measured at 595 nm using a microplate reader (VersaMax™, Molecular Devices, Sunnyvale, CA, USA).

2.2.4. Assessment of Osteoblast Activation after NTP of Specimens Contaminated with Bacteria

The specimens were placed in a multi-well plate, and 1 mL of artificial saliva was dispensed into each well and stirred at room temperature for 2 h to coat specimens. The artificial saliva was aspirated, and the specimens were allowed to dry naturally for 15 min. The liquid culture medium (1.5×10^7 CFU/mL) was dispensed onto each specimen and incubated for 48 h under anaerobic conditions. The specimens on which the bacteria were cultured were treated with NTP for 60 s and 120 s (Figure 2). Immediately after the plasma treatment, osteoblasts (MCT3-E1; 4×10^4 cells/mL) were dispensed onto the specimens and incubated in a CO_2 incubator set at 5% CO_2 and 37 °C. After incubation for 48, 72, and 96 h, osteoblast activation was assessed using a water-soluble tetrazolium salt (WST-8) assay. WST-8 reagent (100 μL; EZ-Cytox, Itsbio, Inc., Seoul, Korea) was dispensed into each well, and the specimens were cultured in an incubator set at 37 °C and 5% CO_2. When the color orange developed into 100 μL of the solution was transferred from each well into a 96-well plate and absorbance was measured at 450 nm with an absorbance meter (ELISA reader: ELx 800UV, Bio-Tek Instrument., Winooski, VT, USA). The contrasting wavelength was set to 630 nm. All specimens were sterilized, and each group was divided into 30 specimens, and 10 specimens were used at each time.

2.2.5. Observation of Adhesion and Morphology of *P. gingivalis* and Osteoblasts

After each incubation period, the medium was removed from each well, and bacteria and cells not adhering to the specimens were carefully washed away with PBS. To obtain SEM images, cells on the specimens were fixed with 2.5% glutaraldehyde and 2% paraformaldehyde as primary fixatives at room temperature for 2 h. After washing with PBS three times (10 min each), the cells were dehydrated in an ethanol gradient (40%, 50%, 60%, 70%, 80%, and 90%) for 5 min at each concentration, and then in 100% ethanol twice (5 min each). The specimens were immersed in hexamethyldisilazane for 15 min and dried on a clean bench for 1–2 h. the dried specimens were coated with platinum for 60 s in a vacuum using a sputter coater (E-1030, Hitachi), and cell adhesion was observed using SEM (FE-SEM S-4700, Hitachi).

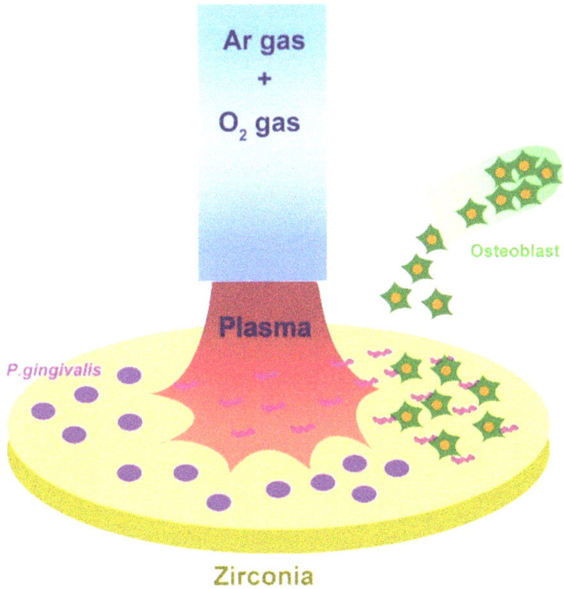

Figure 2. Specimens on which bacteria were cultured and treated with non-thermal plasma.

2.2.6. Statistical Analysis

Statistical analysis was performed using SPSS 21.0 software (SPSS Inc., Chicago, IL, USA). The statistical significance of the crystal violet assay result after 24 h was determined by one-way analysis of variance at a significance level set at $p < 0.05$. Tukey's test was used as a post-hoc test to evaluate the significance of differences between groups. Data from the WST-8 assay after 48, 72, and 96 h did not meet the assumption of normality. The statistical significance of WST-8 assay after 48, 72, and 96 h was tested using Kruskal–Wallis test. The significance level was set at $p < 0.05$. The Mann–Whitney test was used as a post-hoc test to determine the significance between groups, and type I errors were corrected using Bonferroni's method. The significance level was set at $p < 0.017$.

3. Results

3.1. Surface Characteristics

Figure 3 shows that the structure of the zirconia surface did not differ before and after NTP treatment for the different treatment durations. The elemental content on the surface before and after plasma treatment was measured, and changes after treatment were confirmed, using EDX analysis (Figure 4).

A microscopic amount of Ar was detected in the zirconia specimen treated with NTP for more than 300 s. The weight ratio and element ratios for each group are listed in Table 1. The carbon peak decreased in the 60, 300, and 600 s groups following NTP treatment of the zirconia surface compared to in the untreated control group, according to the XPS results. In the carbon 1 s spectrum, the peak was at 28,813.03 counts/s in the control group, whereas the peaks were 12,121.69, 9411.18, and 13,346.04 counts/s in the 60, 300, and 600 s groups, respectively (Table 2).

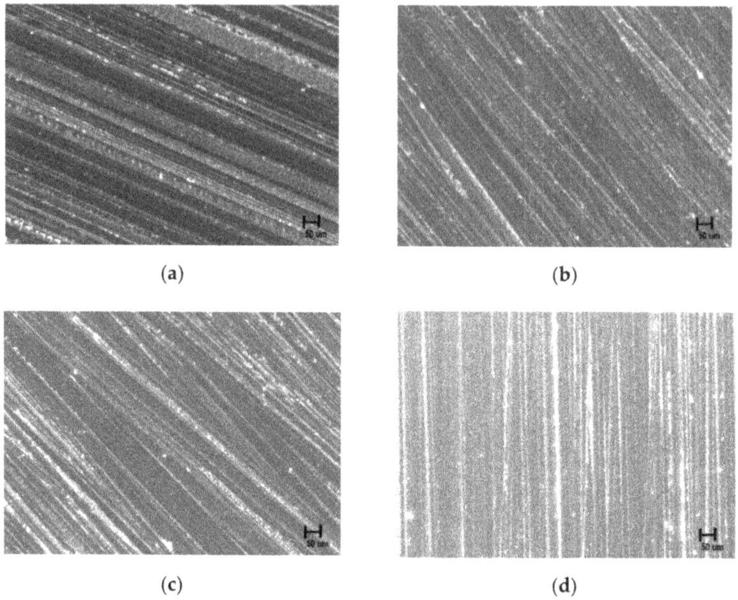

Figure 3. Field emission-scanning electron microscopy surface images (**a**) control group, (**b**) 60 s group, (**c**) 300 s group, (**d**) 600 s group (×3000).

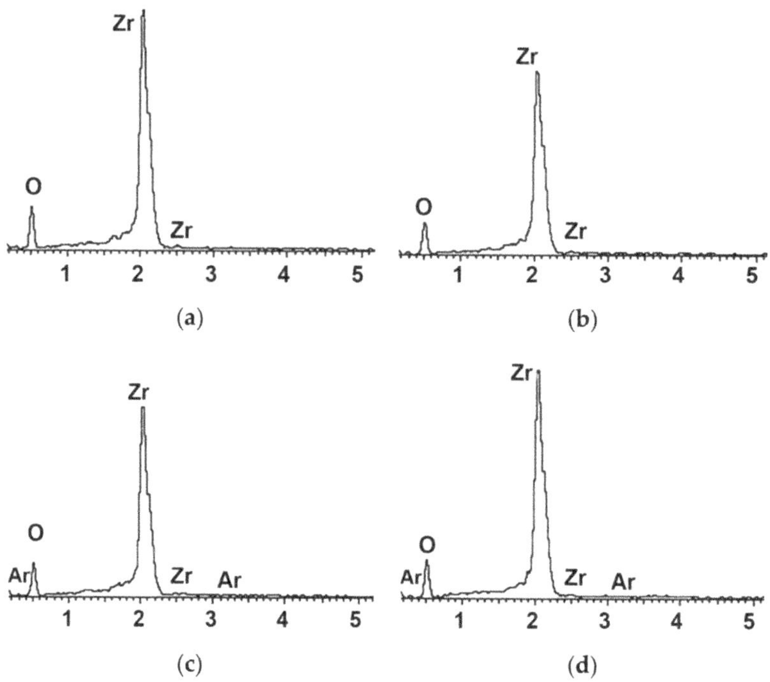

Figure 4. Energy-dispersive X-ray spectroscopy elemental spectrums of the (**a**) control group, (**b**) 60 s group, (**c**) 300 s group, (**d**) 600 s group.

Table 1. Elemental value following plasma treatment.

Group / Element	Control		60 s		300 s		600 s	
	Weight	Atom	Weight	Atom	Weight	Atom	Weight	Atom
O	28.58 *	69.53	27.77	68.67	26.73	67.53	26.90	67.66
Zr	71.42	30.47	72.23	31.33	73.24	32.45	72.95	32.18
Ar	-	-	-	-	0.02	0.02	0.16	0.16

*: %, sum is 100%.

Table 2. X-ray photoelectron spectroscopy C1s & O1s counts/s following plasma treatment.

Group / Element	Control	60 s	300 s	600 s
Carbon (C1s)	28,813 *	12,121	9411	13,346
Oxygen (O1s)	148,403	171,896	129,183	153,686

*: counts/s.

Changes in the oxygen 1 s spectrum slightly increased when plasma was applied to the specimens for 60 s, with no significant difference when plasma was applied for 300 and 600 s. Figure 4 shows the changes in the contact angle of the zirconia surface because of NTP treatment. The contact angle decreased in the NTP-treated groups (60, 120, 180 s group) compared to that in the untreated control group. The average contact angle was 53.83° in the control group and 25.98°, 23.29°, and 21.56° in the 60, 120, and 180 s groups, respectively (Figure 5). Particularly, the contact angle spread immediately after distilled water came into contact with specimens treated with plasma for more than 300 s. The number of specimens used in the experiment was three pieces for each group.

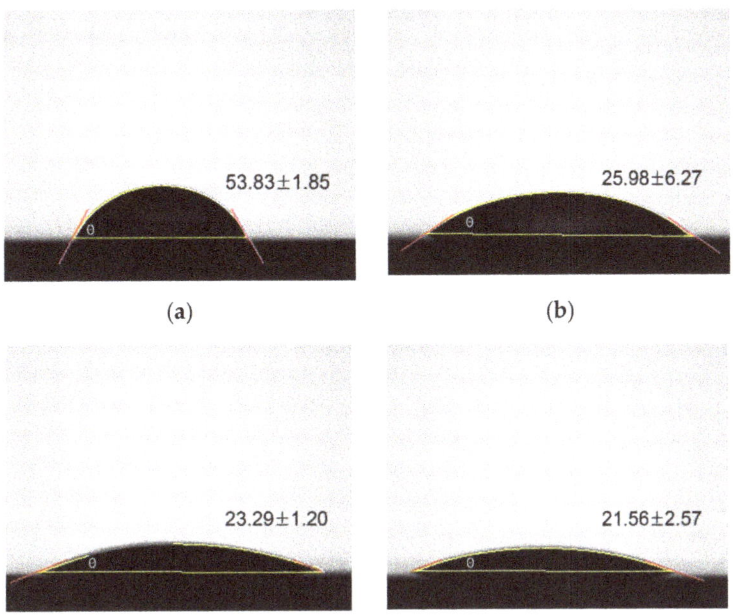

(a) (b) (c) (d)

Figure 5. Water contact angle of specimens. (**a**) control group, (**b**) 60 s group, (**c**) 120 s group, (**d**) 180 s group.

3.2. Biological Response

3.2.1. Bacterial Adhesion after NTP Treatment on Specimens

The group treated with NTP for 60 s showed no significant differences from the control group, whereas *P. gingivalis* adhesion was significantly increased in the groups treated with plasma for 300 and 600 s ($p < 0.05$) (Figure 6).

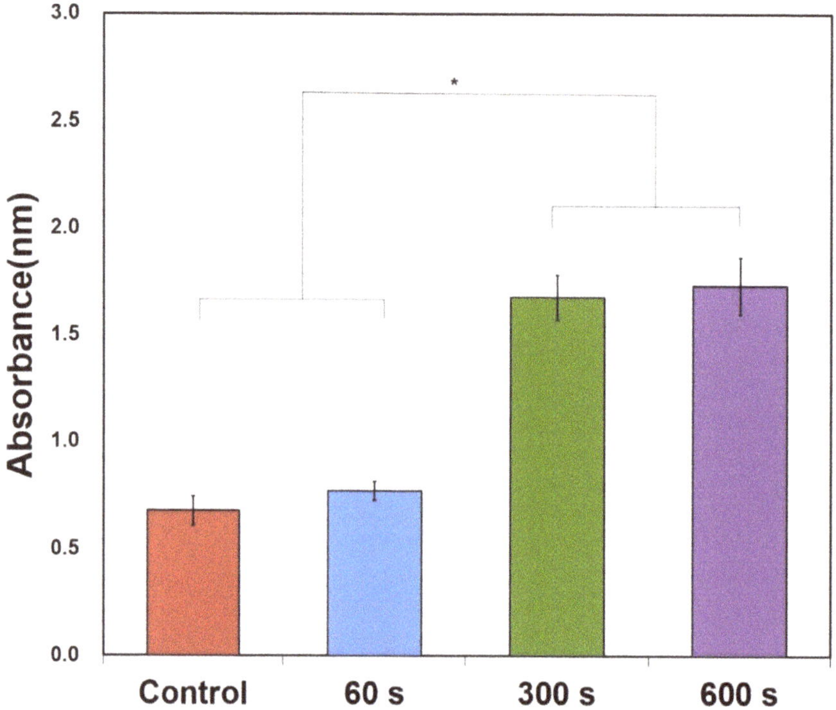

Figure 6. Crystal violet assay after plasma treatment. *: Significant difference at $p < 0.05$.

3.2.2. Osteoblast Activation after NTP Treatment of Bacteria Contaminated Specimens

After 48 h, osteoblast activity was significantly increased in the group treated with plasma for 60 s; after 72 and 96 h, osteoblast activity was significantly increased in all plasma-treated groups (I60 s, I120 s) (Figure 7) ($p < 0.017$). The plasma-treated groups showed approximately two- and four-fold higher activity than the control group (I; inoculated bacteria and no NTP treatment) after 72 and 96 h, respectively.

3.2.3. Microscopic Assessment of *P. gingivalis* and Osteoblast Adhesion

P. gingivalis cultured on plasma-treated specimens was observed using SEM (Figure 8). After 48 h of bacterial culture, *P. gingivalis* was attached to the specimen in the form of bacilli in an oval shape. The shape did not significantly change according to the duration of plasma application, and the size remained similar at approximately 5 μm. *P. gingivalis* was more densely attached to the specimens of the groups treated for 300 and 600 s than to those of the control group and the group treated for 60 s.

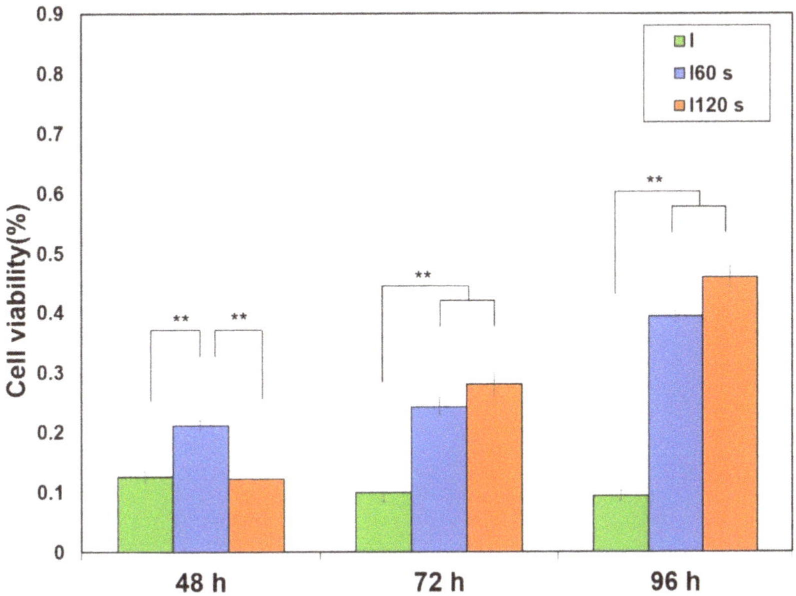

Figure 7. Evaluation of MC3T3-E1 cell viability. **: Significant difference at $p < 0.017$.

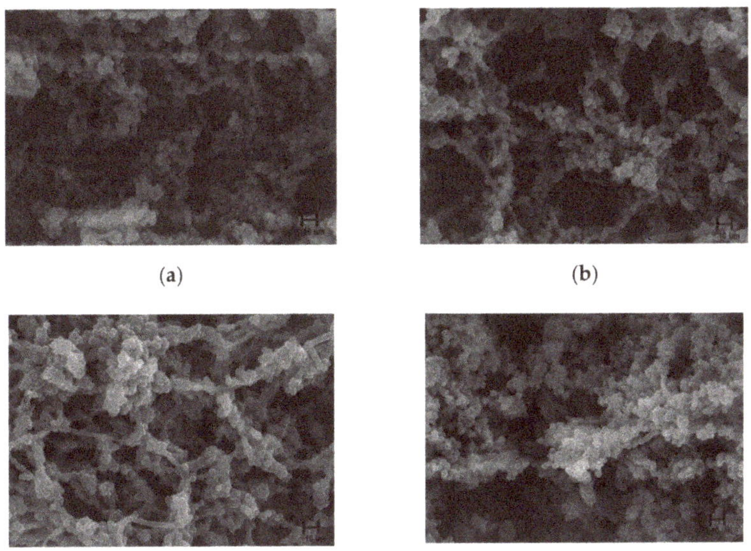

(a) (b) (c) (d)

Figure 8. Field emission-scanning electron microscopy images of *P. gingivalis* cultured for 48 h after non-thermal plasma treatment in the (**a**) control group, (**b**) 60 s group, (**c**) 300 s group, (**d**) 600 s group (×5000).

Zirconia specimens contaminated with *P. gingivalis* were treated with NTP, and the adhesion of osteoblasts was observed using an SEM (Figure 9). More bacteria were destroyed immediately after NTP treatment than in the control group. Additionally, bacterial destruction was more extensive in the 120 s group than in the 60 s group (Figure 9a–c). At

24 h after osteoblast injection, osteoblasts adhered in the NTP treated groups (I60, I120 s group) (Figure 9d–f). On the surface on which the dispensed osteoblasts were incubated for 48 h, osteoblast adhesion was higher in the plasma-treated groups. In the control group, bacteria and cells could not adhere to the specimens and were destroyed (Figure 9g–i). In the plasma-treated groups, osteoblast adhesion on the surface of the 72-h-incubated specimens was higher than that in the 48-h-incubated specimens, whereas, in the control group, more bacteria and cells were destroyed in the specimens incubated for 72 h than in those incubated for 48 h (Figure 9j–l). On the surface of 96 h incubated specimens, osteoblasts were abundantly distributed in the plasma-treated groups, whereas both bacteria and cells died in the control group (Figure 9m–o). Overall, as the incubation period increased, the control group exhibited the death of both osteoblast cells and bacteria, whereas, in the plasma-treated group, osteoblasts were attached to and richly distributed on the surface, and activated.

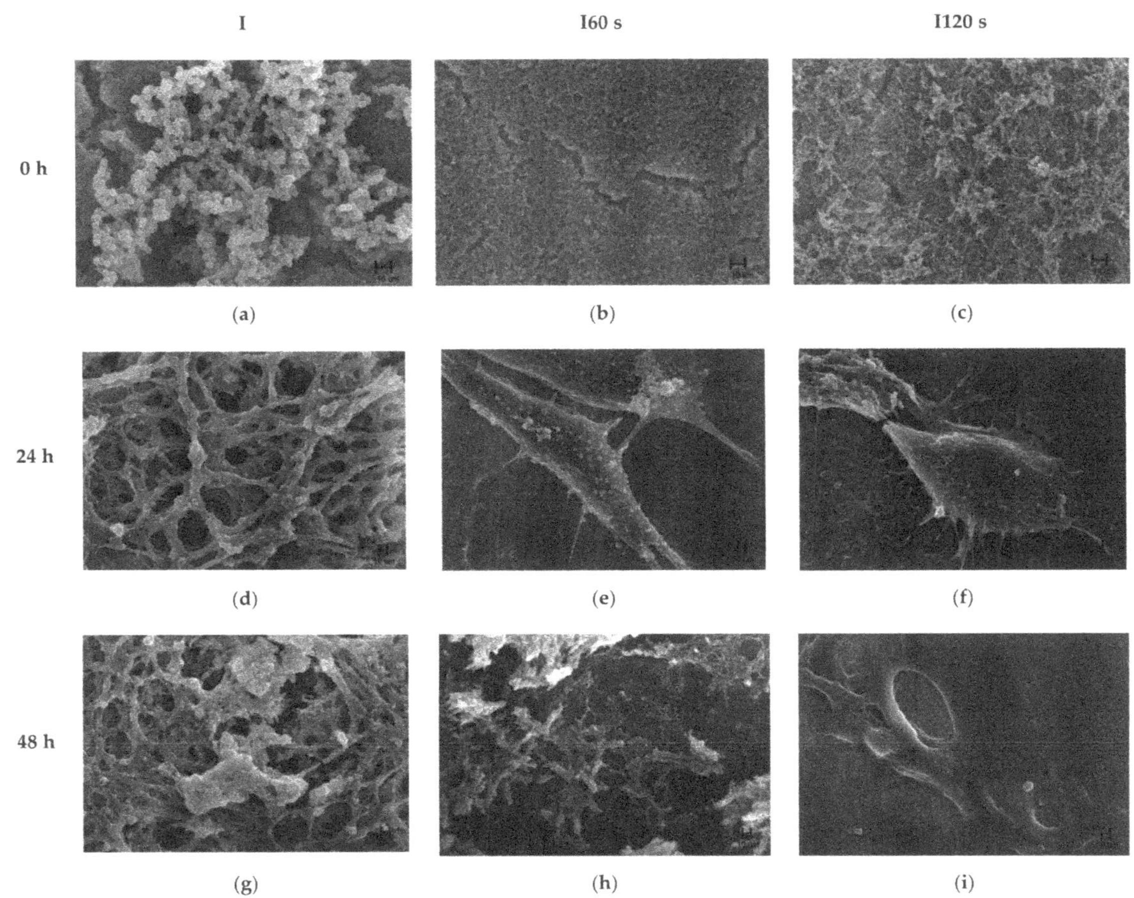

Figure 9. *Cont.*

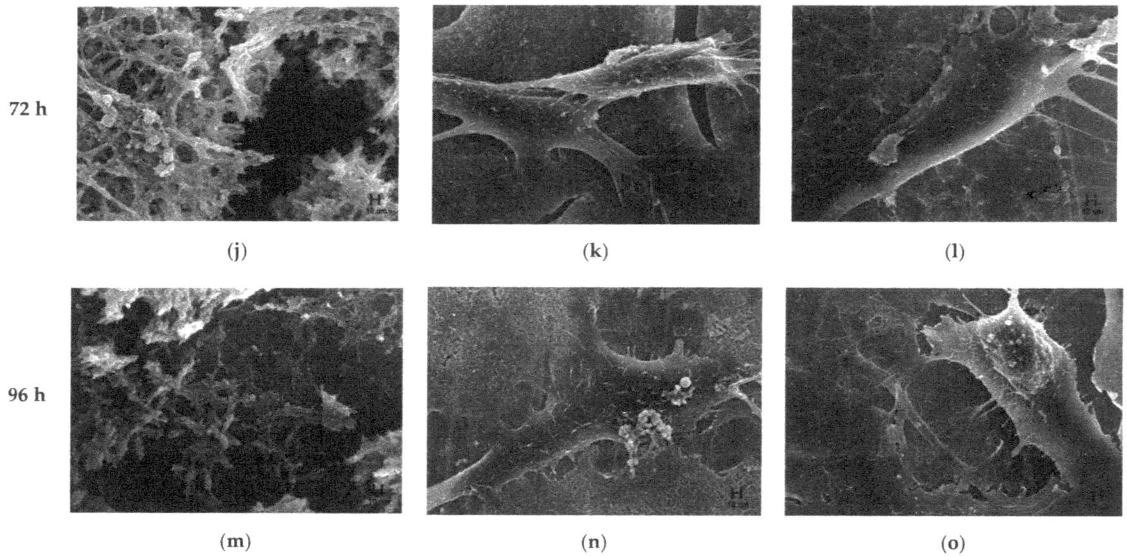

Figure 9. Field emission-scanning electron microscopy images of osteoblastic-like MC3T3-E1 cells cultured after 0 h (**a**–**c**) (×5000), 24 h (**d**–**f**), 48 h (**g**–**i**), 72 h (**j**–**l**), and 96 h (**m**–**o**) (×3000) on contaminated zirconia surfaces.

4. Discussion

Changes in the hydrophilicity of biomaterials promote cell adhesion and growth [25,29]. NTP treatment increases hydrophilicity by altering the chemical composition on the surface of biomaterials [30]. Duske et al. [22] reported that the adhesion and differentiation of osteoblasts showed the greatest increase when plasma was applied to a titanium surface using argon gas containing 1% oxygen as a process gas. The plasma process conditions described by Duske et al. were referenced when formulating our study; in our experiments, the shape of the zirconia surface did not change after NTP treatment, which is consistent with the results of studies in which atmospheric plasma was applied to other surfaces [31]. However, the contact angle with the surface decreased after plasma treatment compared to that on the non-plasma treated surface and further decreased with the treatment duration. We assessed the hydrophilicity of the surface and confirmed the functional groups determining the surface charge. The carbon peak on the surface of zirconia decreased with atmospheric plasma in proportion to the plasma treatment duration, and Ar was detected. Based on the detection of trace Ar when NTP was applied for more than 300 s, changes in the chemical composition of the surface may have increased when plasma was applied for more than 300 s. Previous studies [32,33] reported that after plasma application, the concentrations of CO/CO and O=CO increased and free radicals were generated by the plasma, separating the CC/CH bonds, thereby inducing the generation of OH and O groups, which are related to hydrophilicity. As carbon adversely affects cellular activity, removing carbon from the surface can enhance the adhesion, proliferation, and differentiation of osteoblasts [34,35]. In contrast, bacterial adhesion and differentiation react differently to the hydrophilicity of biomaterials. Bacterial adhesion is more favorably achieved on hydrophobic or nonpolar surfaces than on hydrophilic surfaces [36,37]. In contrast to the findings of Yoo, et al. [21], who reported that zirconia had an antibacterial effect on *Streptococcus mutans* culture after NTP treatment, we found that adhesion of *P. gingivalis* increased after plasma application. Bacterial adhesion is determined by the effects of various interactions between the bacterial film and biomaterial surface, such as hydrogen, ionic, and covalent bonding. Depending on whether the bacterial cell membrane was formed, the difference in adhesion after NTP treatment between *S. mutans* and *P. gingivalis* may be because *S. mutans* is a gram-positive coccus, and *P. gingivalis* is a gram-negative bacillus. Liu et al. [38] suggested that bacteria adhere better to hydrophobic surfaces than to hydrophilic surfaces, whereas Boks et al. [39,40] reported that *Staphylococcus epidermidis* adheres better to hydrophilic surfaces. *S. mutans* has been reported to be hydrophobic [41], and *P. gingivalis* exhibits both hydrophilic and hydrophobic characteristics [42]. Therefore, the increased hydrophilicity due to atmospheric-pressure plasma treatment on the zirconia surface decreased the adhesion of hydrophobic *S. mutans* and increased the adhesion of *P. gingivalis*.

According to Ange et al. [43], the adhesion of bacteria increases as the roughness of the surface increases. Therefore, in this study, in order to reduce the roughness of the zirconia surface, all the specimens were finally polished for 60 s with #800 SiC grit abrasive paper after the specimen was prepared, and a smooth surface was created with the same roughness. Huh et al. [44] report the zirconia surface polished with diamond and SiC for 60 s did not cause any surface changes and was clinically acceptable. There was no change in surface roughness before and after plasma treatment, and there was no physical change in the dimensional properties of the zirconia surface before and after plasma treatment. This is in agreement with the findings of Shon et el. [27]. Additionally, Guo et el. [45] reported that plasma treatment increased osseointegration without changing the zirconia surface structure.

We investigated the activity of osteoblasts after plasma sterilization of zirconia contaminated with *P. gingivalis*. When the duration of plasma treatment was set to be identical to the duration of assessment of *P. gingivalis* adhesion activity, all *P. gingivalis* were destroyed in more than 300 s, leaving only debris; this was considered an unsuitable condition for assessing osteoblast activity after plasma sterilization treatment of zirconia contaminated with *P. gingivalis*. Therefore, considering the clinical use time, the experiment was performed for

60 and 120 s, during which not all *P. gingivalis* were destroyed. The first null hypothesis of this study (i.e., there would be no difference in the degree of adhesion of *P. gingivalis* when zirconia was treated with NTP) was thus rejected. The second null hypothesis, which was that there would be no difference in the degree of adhesion of osteoblasts even if zirconia inoculated with *P. gingivalis* was treated with NTP, was also rejected. The results of the WST-8 assay in which osteoblasts were cultured for 24 h in the specimen not treated with NTP after inoculation with *P. gingivalis* indicated that absorbance was higher than that of the plasma-treated specimens under the same conditions; this can be attributed to the fact that not only osteoblasts but also *P. gingivalis* attached to the specimens reacted with WST-8 reagent [46]. However, in specimens not treated with plasma, after 48 h, *P. gingivalis* did not increase in number and eventually died, and osteoblasts did not exhibit growth. Both the bacteria and osteoblast cells were killed at 72 h. Guo et al. [47] suggested that lipopolysaccharide, a toxic substance continuously secreted from killed bacteria, inhibits osteoblast differentiation, and induces apoptosis. Osteoblast culture after plasma treatment of specimens contaminated with bacteria has not been widely evaluated. Duske et al. [48] incubated osteoblasts obtained after plasma treatment and after physical intervention, on the surface of a titanium disk contaminated with a bacterial film, to compare the two treatment methods. Their results suggested that a combination of the two treatments is the most effective method for treating peri-implantitis. Annunziata et el. [49] reported that the argon plasma technology could be efficiently used to decontaminate/sterilize previously infected titanium implant surfaces.

In this study, osteoblasts existed between sterilized bacterial remnants when culturing osteoblasts after only plasma treatment without physical intervention. Although not accompanied by physical procedures, osteoblast activity in the groups treated with plasma for more than 60 s increased over time compared to that in the untreated control group. Physical intervention is important in the treatment of peri-implantitis but is associated with a risk of infection and bleeding during surgical procedures. Therefore, when oral implants are exposed because of advanced peri-implantitis, plasma treatment alone can increase osteoblast adhesion and disinfection efficacy in patients who cannot undergo physical procedures because of various conditions including systemic diseases. As a limitation of this study, only *P. gingivalis* was tested among various bacteria causing peri-implantitis, and additional research on other peri-implantitis-causing bacteria is needed.

5. Conclusions

We investigated osteoblast activity after plasma sterilization of zirconia contaminated with *P. gingivalis* and determined the optimal plasma treatment duration for obtaining clinically favorable outcomes in terms of both bacterial contamination and osteoblast activity. We determined that although treatment with NTP for 60 s did not increase the adhesion of *P. gingivalis*, it increased the adhesion of osteoblasts and resulted in sterilization effects. However, when *P. gingivalis* was cultured after 60 s of NTP treatment of zirconia, there was no significant difference from the control group ($p = 0.91$), but the adhesion was significantly increased in the group treated with plasma for 300 s or longer (300 and 600 s groups) ($p < 0.05$). Thus, treatment with NTP for more than 60 s may lead to an increase in bacterial adhesion rather than antibacterial effects. In the treatment of inflammation around zirconia, using NTP maybe a beneficial alternative for patients who cannot be treated with physical intervention due to conditions such as systemic diseases. However, among the various bacteria that cause peri-implantitis, its use is limited to *P. gingivalis*, and additional research is needed on the treatment of inflammation and osseointegration of NTP on other dental materials such as titanium materials contaminated with other types of bacteria.

Author Contributions: Conceptualization, S.-K.L., M.-K.J., Y.-J.J. and H.-P.L.; methodology, H.C. and H.-P.L.; software, S.-K.L., M.-K.J. and Y.-J.J.; validation, C.P. and H.C.; formal analysis, M.-K.J. and C.P.; investigation, S.-K.L., M.-K.J. and Y.-J.J.; resources, H.C. and H.-P.L.; data curation, H.C. and H.-P.L.; writing—original draft preparation, S.-K.L., M.-K.J. and Y.-J.J.; writing—review and editing, S.-K.L. and M.-K.J.; visualization, S.-K.L., M.-K.J. and Y.-J.J.; supervision, C.P.; project administration,

H.C. and H.-P.L.; funding acquisition, H.-P.L. All authors have read and agreed to the published version of the manuscript.

Funding: This work was funded by the National Research Foundation of Korea (NRF) grant funded by the Korean government (MSIT) (No. NRF-2020R1C1C1005683).

Institutional Review Board Statement: Not applicable.

Informed Consent Statement: Not applicable.

Data Availability Statement: Not applicable.

Acknowledgments: This work was supported by the National Research Foundation of Korea (NRF) grant funded by the Korean government (MSIT) (No. NRF-2020R1C1C1005683).

Conflicts of Interest: The authors declare no conflict of interest.

References

1. Jemat, A.; Ghazali, M.J.; Razali, M.; Otsuka, Y. Surface modifications and their effects on titanium dental implants. *BioMed Res. Int.* **2015**, *2015*, 791725. [CrossRef] [PubMed]
2. Kim, K.T.; Eo, M.Y.; Nguyen, T.T.H.; Kim, S.M. General review of titanium toxicity. *Int. J. Implant. Dent.* **2019**, *5*, 10. [CrossRef] [PubMed]
3. Sicilia, A.; Cuesta, S.; Coma, G.; Arregui, I.; Guisasola, C.; Ruiz, E.; Maestro, A. Titanium allergy in dental implant patients: A clinical study on 1500 consecutive patients. *Clin. Oral Implants Res.* **2008**, *19*, 823–835. [CrossRef] [PubMed]
4. Siddiqi, A.; Payne, A.G.T.; De Silva, R.K.; Duncan, W.J. Titanium allergy: Could it affect dental implant integration? *Clin. Oral Implants Res.* **2011**, *22*, 673–680. [CrossRef] [PubMed]
5. Yoshinari, M. Future prospects of zirconia for oral implants—A review. *Dent. Mater. J.* **2020**, *39*, 37–45. [CrossRef] [PubMed]
6. Sivaraman, K.; Chopra, A.; Narayan, A.I.; Balakrishnan, D. Is zirconia a viable alternative to titanium for oral implant? A critical review. *J. Prosthodont. Res.* **2018**, *62*, 121–133. [CrossRef] [PubMed]
7. Hanawa, T. Zirconia versus titanium in dentistry: A review. *Dent. Mater. J.* **2020**, *39*, 24–36. [CrossRef]
8. Hu, J.; Atsuta, I.; Ayukawa, Y.; Zhou, T.; Narimatsu, I.; Koyano, K. Effect of titanium or zirconia implant abutments on epithelial attachments after ultrasonic cleaning. *J. Oral Sci.* **2020**, *62*, 331–334. [CrossRef] [PubMed]
9. Hafezeqoran, A.; Koodaryan, R. Effect of zirconia dental implant surfaces on bone Integration: A systematic review and meta-analysis. *BioMed Res. Int.* **2017**, *2017*, 9246721. [CrossRef]
10. Sanz-Martin, I.; Sanz-Sanchez, I.; Carrillo de Albornoz, A.; Figuero, E.; Sanz, M. Effects of modified abutment characteristics on peri-implant soft tissue health: A systematic review and meta-analysis. *Clin. Oral Implants Res.* **2018**, *29*, 118–129. [CrossRef]
11. Roehling, S.; Astasov-Frauenhoffer, M.; Hauser-Gerspach, I.; Braissant, O.; Woelfler, H.; Waltimo, T.; Kniha, H.; Gahlert, M. In vitro biofilm formation on titanium and zirconia implant surfaces. *J. Periodontol.* **2017**, *88*, 298–307. [CrossRef] [PubMed]
12. Ganbold, B.; Kim, S.K.; Heo, S.J.; Koak, J.Y.; Lee, Z.H.; Cho, J. Osteoclastogenesis behavior of zirconia for dental implant. *Materials* **2019**, *12*, 732. [CrossRef] [PubMed]
13. Daubert, D.M.; Weinstein, B.F. Biofilm as a risk factor in implant treatment. *Periodontology* **2019**, *81*, 29–40. [CrossRef] [PubMed]
14. Da Silva, E.S.; Feres, M.; Figueiredo, L.C.; Shibli, J.A.; Ramiro, F.S.; Faveri, M. Microbio logical diversity of peri-implantitis biofilm by Sanger sequencing. *Clin. Oral Implants Res.* **2014**, *25*, 1192–1199. [CrossRef] [PubMed]
15. Mohanty, R.; Asopa, S.J.; Joseph, M.D.; Singh, B.; Rajguru, J.P.; Saidath, K.; Sharma, U. Red complex: Polymicrobial conglomerate in oral flora: A review. *J. Fam. Med. Prim. Care* **2019**, *8*, 3480–3486. [CrossRef]
16. Thurnheer, T.; Belibasakis, G.N. Incorporation of staphylococci into titanium-grown biofilms: An in vitro "submucosal" biofilm model for peri-implantitis. *Clin. Oral Implants Res.* **2016**, *27*, 890–895. [CrossRef]
17. Zhou, X.; Wu, D.; Liang, D.; Zhang, W.; Shi, Q.; Cao, Y. Evaluation of modified cold-atmospheric pressure plasma (MCAP) for the treatment of peri-implantitis in beagles. *Oral Dis.* **2020**, *28*, 495–502. [CrossRef]
18. Yang, Y.; Guo, J.; Zhou, X.; Liu, Z.; Wang, C.; Wang, K.; Zhang, J.; Wang, Z. A novel cold atmospheric pressure air plasma jet for peri-implantitis treatment: An in vitro study. *Dent. Mater. J.* **2018**, *37*, 157–166. [CrossRef]
19. Heinlin, J.; Morfill, G.; Landthaler, M.; Stolz, W.; Isbary, G.; Zimmermann, J.L.; Shimizu, T.; Karrer, S. Plasma medicine: Possible applications in dermatology. *J. Dtsch. Derm. Ges.* **2010**, *8*, 968–976. [CrossRef]
20. Duarte, S.; Panariello, B.H.D. Comprehensive biomedical applications of low temperature plasmas. *Arch. Biochem. Biophys.* **2020**, *693*, 108560. [CrossRef]
21. Yoo, E.M.; Uhm, S.H.; Kwon, J.S.; Choi, H.S.; Choi, E.H.; Kim, K.M.; Kim, K.N. The study on inhibition of planktonic bacterial growth by non-thermal atmospheric pressure plasma jet treated surfaces for dental application. *J. Biomed. Nanotechnol.* **2015**, *11*, 334–341. [CrossRef] [PubMed]
22. Duske, K.; Koban, I.; Kindel, E.; Schröder, K.; Nebe, B.; Holtfreter, B.; Jablonowski, L.; Weltmann, K.D.; Kocher, T. Atmospheric plasma enhances wettability and cell spreading on dental implant metals. *J. Clin. Periodontol.* **2012**, *39*, 400–407. [CrossRef] [PubMed]

23. Canullo, L.; Micarelli, C.; Bettazzoni, L.; Magnelli, A.; Baldissara, P. Shear bond strength of veneering porcelain to zirconia after argon plasma treatment. *Int. J. Prosthodont.* **2014**, *27*, 137–139. [CrossRef] [PubMed]
24. Valverde, G.B.; Coelho, P.G.; Janal, M.N.; Lorenzoni, F.C.; Carvalho, R.M.; Thompson, V.P.; Weltemann, K.D.; Silva, N.R. Surface characterisation and bonding of Y-TZP following non-thermal plasma treatment. *J. Dent.* **2013**, *41*, 51–59. [CrossRef] [PubMed]
25. Watanabe, H.; Saito, K.; Kokubun, K.; Sasaki, H.; Yoshinari, M. Change in surface properties of zirconia and initial attachment of osteoblastlike cells with hydrophilic treatment. *Dent. Mater. J.* **2012**, *31*, 806–814. [CrossRef]
26. Park, Y.S.; Chung, S.H.; Shon, W.J. Peri-implant bone formation and surface characteristics of rough surface zirconia implants manufactured by powder injection molding technique in rabbit tibiae. *Clin. Oral Implants Res.* **2013**, *24*, 586–591. [CrossRef] [PubMed]
27. Shon, W.J.; Chung, S.H.; Kim, H.K.; Han, G.J.; Cho, B.H.; Park, Y.S. Peri-implant bone formation of non-thermal atmospheric pressure plasma-treated zirconia implants with different surface roughness in rabbit tibiae. *Clin. Oral Implants Res.* **2014**, *25*, 573–579. [CrossRef] [PubMed]
28. Li, Y.H.; Tang, N.; Aspiras, M.B.; Lau, P.C.; Lee, J.H.; Ellen, R.P.; Cvitkovitch, D.G. A quorum-sensing signaling system essential for genetic competence in Streptococcus mutans is involved in biofilm formation. *J. Bacteriol.* **2002**, *184*, 2699–2708. [CrossRef] [PubMed]
29. Zheng, M.; Yang, Y.; Liu, X.Q.; Liu, M.Y.; Zhang, X.F.; Wang, X.; Li, H.P.; Tan, J.G. Enhanced biological behavior of in vitro human gingival fibroblasts on cold plasma-treated zirconia. *PLoS ONE* **2015**, *10*, e0140278. [CrossRef] [PubMed]
30. Bazaka, K.; Jacob, M.V.; Crawford, R.J.; Ivanova, E.P. Plasma-assisted surface modification of organic biopolymers to prevent bacterial attachment. *Acta Biomater.* **2011**, *7*, 2015–2028. [CrossRef]
31. Pan, Y.H.; Lin, J.C.Y.; Chen, M.K.; Salamanca, E.; Choy, C.S.; Tsai, P.Y.; Leu, S.J.; Yang, K.C.; Huang, H.M.; Yao, W.L.; et al. Glow discharge plasma treatment on zirconia surface to enhance osteoblastic-Like cell differentiation and antimicrobial effects. *Materials* **2020**, *13*, 3771. [CrossRef] [PubMed]
32. Wang, R.; Shen, Y.; Zhang, C.; Yan, P.; Shao, T. Comparison between helium and argon plasma jets on improving the hydrophilic property of PMMA surface. *Appl. Sur. Sci.* **2016**, *367*, 401–406. [CrossRef]
33. Kim, M.C.; Song, D.K.; Shin, H.S.; Baeg, S.H.; Kim, G.S.; Boo, J.H.; Han, J.G.; Yang, S.H. Surface modification for hydrophilic property of stainless steel treated by atmospheric-pressure plasma jet. *Surf. Coat. Technol.* **2003**, *171*, 312–316. [CrossRef]
34. Aita, H.; Hori, N.; Takeuchi, M.; Suzuki, T.; Yamada, M.; Anpo, M.; Ogawa, T. The effect of ultraviolet functionalization of titanium on integration with bone. *Biomaterials* **2009**, *30*, 1015–1025. [CrossRef] [PubMed]
35. Zheng, Z.; Ao, X.; Xie, P.; Wu, J.; Dong, Y.; Yu, D.; Wang, J.; Zhu, Z.; Xu, H.H.K.; Chen, W. Effects of novel non-thermal atmospheric plasma treatment of titanium on physical and biological improvements and in vivo osseointegration in rats. *Sci. Rep.* **2020**, *10*, 10637. [CrossRef]
36. Krasowska, A.; Sigler, K. How microorganisms use hydrophobicity and what does this mean for human needs? *Front. Cell Infect. Microbiol* **2014**, *4*, 112. [CrossRef]
37. Mazumder, S.; Falkinham, J.O., 3rd; Dietrich, A.M.; Puri, I.K. Role of hydrophobicity in bacterial adherence to carbon nanostructuRes. and biofilm formation. *Biofouling* **2010**, *26*, 333–339. [CrossRef]
38. Liu, Y.; Yang, S.F.; Li, Y.; Xu, H.; Qin, L.; Tay, J.H. The influence of cell and substratum surface hydrophobicities on microbial attachment. *J. Biotechnol.* **2004**, *110*, 251–256. [CrossRef]
39. Boks, N.P.; Busscher, H.J.; van der Mei, H.C.; Norde, W. Bond-strengthening in staphylococcal adhesion to hydrophilic and hydrophobic surfaces using atomic force microscopy. *Langmuir* **2008**, *24*, 12990–12994. [CrossRef] [PubMed]
40. Boks, N.P.; Kaper, H.J.; Norde, W.; van der Mei, H.C.; Busscher, H.J. Mobile and immobile adhesion of staphylococcal strains to hydrophilic and hydrophobic surfaces. *J. Colloid Interface Sci.* **2009**, *331*, 60–64. [CrossRef]
41. Nostro, A.; Cannatelli, M.A.; Crisafi, G.; Musolino, A.D.; Procopio, F.; Alonzo, V. Modifications of hydrophobicity, in vitro adherence and cellular aggregation of Streptococcus mutans by Helichrysum italicum extract. *Lett. Appl. Microbiol.* **2004**, *38*, 423–427. [CrossRef] [PubMed]
42. Kerosuo, E.; Haapasalo, M.; Alli, K.; Lounatmaa, K. Ingestion of Bacteroides buccae, Bacteroides oris, Porphyromonas gingivalis, and Fusobacterium nucleatum by human polymorphonuclear leukocytes in vitro. *Oral Microbiol. Immunol.* **1990**, *5*, 202–207. [CrossRef]
43. Ange, L.; Yan, G.; Tan, J.; Xichun, L.; Quanren, Z.; Zhentao, S. Effects of surface roughness and texture on the bacterial adhesion on the bearing surface of bio-ceramic joInt. implants: An in vitro study. *Ceram. Int.* **2020**, *46*, 6550–6559. [CrossRef]
44. Huh, Y.H.; Park, C.J.; Cho, L.R. Evaluation of various polishing systems and the phase transformation of monolithic zirconia. *J. Prosthet. Dent.* **2016**, *116*, 440–449. [CrossRef] [PubMed]
45. Guo, S.; Liu, N.; Liu, K.; Li, Y.; Zhang, W.; Zhu, B.; Gu, B.; Wen, N. Effects of carbon and nitrogen plasma immersion ion implantation on bioactivity of zirconia. *RSC Adv.* **2020**, *10*, 35917–35929. [CrossRef] [PubMed]
46. Pérez, L.M.; Alvarez, B.L.; Codony, F.; Fittipaldi, M.; Adrados, B.; Peñuela, G.; Morató, J. A new microtitre plate screening method for evaluating the viability of aerobic respiring bacteria in high surface biofilms. *Lett. Appl. Microbiol.* **2010**, *51*, 331–337. [CrossRef]
47. Guo, C.; Yuan, L.; Wang, J.G.; Wang, F.; Yang, X.K.; Zhang, F.H.; Song, J.L.; Ma, X.Y.; Cheng, Q.; Song, G.H. Lipopolysaccharide (LPS) induces the apoptosis and inhibits osteoblast differentiation through JNK pathway in MC3T3-E1 cells. *Inflammation* **2014**, *37*, 621–631. [CrossRef] [PubMed]

48. Duske, K.; Jablonowski, L.; Koban, I.; Matthes, R.; Holtfreter, B.; Sckell, A.; Nebe, J.B.; von Woedtke, T.; Weltmann, K.D.; Kocher, T. Cold atmospheric plasma in combination with mechanical treatment improves osteoblast growth on biofilm covered titanium discs. *Biomaterials* **2015**, *52*, 327–334. [CrossRef] [PubMed]
49. Annunziata, M.; Canullo, L.; Donnarumma, G.; Caputo, P.; Nastri, L.; Guida, L. Bacterial inactivation/sterilization by argon plasma treatment on contaminated titanium implant surfaces: In vitro study. *Med. Oral Patol. Oral Cir. Bucal* **2016**, *21*, e118. [CrossRef]

materials

MDPI

Article

Human Gingival Fibroblast Adhesion and Proliferation on Hydroxyapatite-Coated Zirconia Abutment Surfaces

Oskar Bunz [1,*], Marie-Christine Steegmann [2], Korbinian Benz [3], Holger Testrich [4], Antje Quade [4], Ella A. Naumova [5], Wolfgang H. Arnold [5], Katja Fricke [4], Andree Piwowarczyk [1,†] and Thomas Dittmar [6,†]

1 Department of Prosthodontics, School of Dentistry, Faculty of Health, Witten/Herdecke University, 58455 Witten, Germany; andree.piwowarczyk@uni-wh.de
2 Independent Researcher, 55116 Mainz, Germany; mcsteegmann@googlemail.com
3 Department of Oral Surgery and Dental Emergency Care, Faculty of Health, Witten/Herdecke University, 58455 Witten, Germany; korbinian.benz@uni-wh.de
4 Research Group of Bioactive Surfaces, INP Greifswald e.V., 17489 Greifswald, Germany; holger.testrich@inp-greifswald.de (H.T.); quade@inp-greifswald.de (A.Q.); fricke@nebula-biocides.de (K.F.)
5 Department of Biological and Material Sciences in Dentistry, School of Dentistry, Faculty of Health, Witten/Herdecke University, 58455 Witten, Germany; ella.naumova@uni-wh.de (E.A.N.); wolfgang.arnold@uni-wh.de (W.H.A.)
6 Institute of Immunology, Center for Biomedical Education and Research (ZBAF), Witten/Herdecke University, 58453 Witten, Germany; thomas.dittmar@uni-wh.de
* Correspondence: oskar.bunz@uni-wh.de
† These authors contributed equally to this work.

Citation: Bunz, O.; Steegmann, M.-C.; Benz, K.; Testrich, H.; Quade, A.; Naumova, E.A.; Arnold, W.H.; Fricke, K.; Piwowarczyk, A.; Dittmar, T. Human Gingival Fibroblast Adhesion and Proliferation on Hydroxyapatite-Coated Zirconia Abutment Surfaces. *Materials* **2022**, *15*, 3625. https://doi.org/10.3390/ma15103625

Academic Editors: Antonio Scarano and Gabriele Cervino

Received: 11 February 2022
Accepted: 16 May 2022
Published: 19 May 2022

Publisher's Note: MDPI stays neutral with regard to jurisdictional claims in published maps and institutional affiliations.

Abstract: Applying antibacterial coatings to dental implant materials seems reasonable but can have negative influences on desired cell adhesion and healing. In this study, zirconia abutment specimens interacting with gingival tissue were used. The aim was to compare the influence of machined or coated zirconia surfaces on the adhesion and proliferation of human gingival fibroblasts (HGF-1). Surface modifications were performed using atmospheric plasma coating with hydroxyapatite, zinc, and copper. Zirconia specimens were divided into four groups: hydroxyapatite, hydroxyapatite with zinc oxide (ZnO), hydroxyapatite with copper (Cu), and an untreated machined surface. After the characterization of the surface conditions, the morphology of adhered HGF-1 was determined by fluorescence staining and subjected to statistical evaluation. The visual analysis of cell morphology by SEM showed flat, polygonal, and largely adherent fibroblast cells in the untreated group, while round to partially flat cells were recorded in the groups with hydroxyapatite, hydroxyapatite + ZnO, and hydroxyapatite + Cu. The cell membranes in the hydroxyapatite + ZnO and hydroxyapatite + Cu groups appeared porous. The results show that HGF-1 adhere and proliferate well on machined zirconia, while plasma coating with hydroxyapatite or hydroxyapatite mixtures does not lead to increased adhesion or proliferation.

Keywords: ceramic implant; atmospheric plasma spraying; abutment; human gingival fibroblasts

1. Introduction

A dental implant can be used to replace a tooth, including the root, by screwing an artificial root replacement into the jawbone. Through Branemark et al., implantology has become an integral part of dentistry [1]. With increasing experience and the further development of implant surfaces, the scientific focus has changed from investigating osseointegration to investigating the peri-implant soft tissue.

Accumulations of plaque and bacteria on implant surfaces are known to lead to the inflammation of the peri-implant or surrounding tissue as well as the gingiva, thus contributing to treatment failure [2,3]. In numerous studies, the bacterial plaque biofilm has been shown to be a crucial factor in the development of peri-implant mucositis [4]. It is this plaque biofilm that triggers an initial immune response in the host soft tissue, comparable to that on natural teeth [5].

In peri-implant mucositis, plaque accumulation is the main etiologic factor, and peri-implantitis is also plaque-associated, making it clear that insufficient peri-implant soft tissue facilitates submucosal plaque accumulation or implant/abutment surface contamination [6]. Thus, an immobile keratinized mucosa around the dental implant can protect against bone loss.

If the composition of gingival tissue on a healthy tooth is compared with that on a dental implant, similarities and minor differences become apparent. The differences concern the percentage distribution of the tissue types as well as the morphology and organization of the cells [7], and these differences result from the absence of root cement. Instead of a direct attachment, there is more of an adhesion of the supracrestal connective tissue. These fibers also do not run vertically into the root cementum as in the natural tooth, but mainly run parallel to the abutment surface [7]. Additionally, the gingiva around natural teeth has a higher proportion of blood vessels than the peri-implant soft tissue [8].

Fibroblasts appear to have less contact with the implant surface than with natural teeth. This seems to be limited to fibroblasts only, as collagen fibers, for example, seem to occur normally on implant surfaces [9].

Knowing that the main factor of peri-implant inflammation is bacterial adhesion, many techniques of antibacterial treatment have been developed. For example, there are various techniques for mechanical debridement using ultrasound, air abrasion, or laser or antibiotic therapy [10]. Because these techniques do not always work, there are considerations made to create an antibacterial implant surface. One technique for coating implants is atmospheric plasma spraying (APS), which is the most widely used technique for coating implants among thermal spraying methods [11,12]. Plasma coatings of hydroxyapatite on (metallic) implants have been used in dentistry and orthopedics since the 1980s [13]. To date, this process has mostly been used on metallic substrates, with a few also considering other materials, such as PEEK or composites [14,15]. In most cases, this technique is used to create a more attractive surface for better implant healing. However, it is also possible to apply antibacterial elements on the surface [16].

The purpose of this study was to investigate whether treating zirconia surfaces with an additional coating of hydroxyapatite and antibacterial substances leads to a surface that is attractive to fibroblasts. The surface properties after coating and the influence of machined or coated yttrium-stabilized zirconia surfaces on the adhesion and proliferation of human gingival fibroblasts were analyzed.

2. Materials and Methods

2.1. CAD/CAM Fabrication of Zirconia Specimens

The specimens were provided by VITA Zahnfabrik (Bad Säckingen, Germany). They were made of yttrium-stabilized zirconia (3-YSZ), which is used as a one-piece, cylindroconical screw ceramic implant (ceramic.implant, vitaclinical; VITA Zahnfabrik, Bad Säckingen, Germany) [17].

The specimens were provided as circular discs with a diameter of 13 mm and a height of 2 mm. They all had a machined surface when supplied.

2.2. Surface Treatment

The modification of the sample surfaces was carried out at the Leibniz Institute for Plasma Science and Technology e.V. (INP, Greifswald, Germany) using plasma spray coating. In this process, ceramics or metals are partially or completely melted by means of a thermal plasma and deposited onto surfaces at speeds of up to 450 m/s. The coating was applied using the plasma spray torch Oerlikon Metco MultiCoat, F4MB-XL (Darmstadt, Germany). The powders used were "Metco 6902" (hydroxyapatite), "Metco 55" (copper) (Oerlikon metco, Kelsterbach, Germany), and zinc oxide (>99% purity) (Carl Roth GmbH + Co. KG, Karlsruhe, Germany). One sample was sprayed with pure hydroxyapatite, while hydroxyapatite was mixed with 3 wt% copper and 3 wt% zinc oxide for the other two sets of samples, respectively, according to a previous publication [18]. To obtain a uniform mixing of the powders, the powder mixture was treated for 2 h in a rotary mixer. The particle size of the powder varied between 0.5 and 180 µm (Table 1). The plasma spray process was carried out at a spray distance of 8.5 cm with a gas mixture of 40 slm argon and an admixture of 5 slm nitrogen. The spraying procedure was repeated until a dense layer was produced (19.75 µm and 36.49 µm were reached (Table 1)).

Table 1. On the left: Particle sizes of the powder hydroxyapatite (HAp), copper (Cu), and zinc oxide (ZnO) used. The median particle size ($d_{(p)}$) and range are listed in µm. On the right: The coating thicknesses of the surfaces after the coating procedure are shown in µm.

Particle Sizes of the Powder			Coating Thickness	
Powder	$d_{(p)}$ (µm)	Range (µm)	Powder	Thickness (µm)
HAp	93.78	20–180	HAp	36.49
Cu	67.79	20–130	HAp + 3wt% Cu	32.94
ZnO	2.60	0.5–50	HAp + 3 wt% ZnO	19.75

2.3. Surface Characterization

X-ray photoelectron spectroscopy (XPS) measurements were carried out to determine the composition of the coatings (Axis Supra DLD, Kratos, Manchester, UK). For this purpose, line scans with a distance of 0.5 mm were made over the coatings.

The surface roughness was investigated with the optical measuring system Alicona InfiniteFocus and the computer software Alicona IFM 3.2 (Alicona Imagine GmbH, Raaba/Graz, Austria). For this purpose, the test specimens were aligned at an angle of 90° below the objective lens to ensure the most direct possible view of the surface.

Twelve images (each 508.86 × 407.09 µm) of each group were recorded at 20× magnification. Each of the images was computer analyzed at three different measurement areas (100 × 100 µm) with respect to the surface texture Sa (in µm).

The data were checked for normal distribution using the Kolmogorov–Smirnov test (GraphPad Prism, Version 9.0.1, GraphPad Software, La Jolla, CA, USA). Since the data were not normally distributed, significance was determined using the Kruskal–Wallis test and post hoc test via Dunn's multiple comparisons (alpha level 0.05).

2.4. Cell Culture

The specimens were placed into 24-well plates using sterilized forceps, each without touching the surface, and disinfected in 70% ethanol for 20 min. Subsequently, the test specimens were washed three times with phosphate-buffered saline (PBS) (PAN-Biotech GmbH, Aidenbach, Germany) for 5 min each.

Human gingival fibroblasts (HGF-1, LOT 70001246, ATCC, Manassas, VA, USA) were selected for study. HGF-1 cells were cultured in Dulbecco's Modified Eagle Medium (DMEM) (Sigma-Aldrich, St. Louis, MO, USA) supplemented with 10% Fetal Calf Serum (FCS) (c.c.pro, Oberdorla, Germany), 1% penicillin/streptomycin (PAN-Biotech GmbH, Aidenbach, Germany), and Fibroblast Growth Factor (FGF) (100 µg/mL, Sigma-Aldrich, St. Louis, MO, USA) at 37 °C and 5% CO_2 in a humidified atmosphere incubator. The medium was changed daily, as this had a positive effect on cell growth. The sufficient proliferation of fibroblasts was evident in passages five to seven, and only these passages were used for the experiments.

To detect HGF-1 fibroblasts on the specimen's surface, the cells were stained with 5-chloromethylfluorescein diacetate CMFDA (CellTracker^{TM} Green; ThermoFisher Scientific, Waltham, USA). For this purpose, cells were incubated in 5 µM CellTracker^{TM} Green for 45 min at 37 °C and then centrifuged and washed once with PBS (PAN-Biotech GmbH,

Aidenbach, Germany). Then, 15,000 cells per specimen were seeded and cultured either for 24 h or for 72 h.

2.5. Confocal Laser Scanning Microscopy

To determine the adhesion and proliferation of HGF-1 on the surfaces of specimens, images were obtained byusing a Leica TCS SP5 confocal laser scanning microscope (Leica, Wetzlar, Germany) from all test groups (n = 12/group) at time point 1 (after 24-h cultivation) and from all test groups at time point 2 (n = 6/group; after 24 h and 72 h of cultivation).

The test specimens were embedded in culture medium on a chamber slide. Sixteen images were taken of each specimen, with the edge length of the captured image being approximately 180 µm. The images/data were analyzed using the ImageJ software (https://imagej.nih.gov/ij/download.html, accessed on 11 February 2022). The mean value was then determined for each test specimen and statistically evaluated. Normal distribution was checked using the Kolmogorov–Smirnov test (GraphPad Prism, Version 9.0.1, GraphPad Software, La Jolla, CA, USA). As the data were normally distributed, one-way ANOVA was used for statistical evaluation as compared to the untreated control, and the significance level was determined to be $p \leq 0.05$.

2.6. Scanning Electron Microscopy

The morphological examination of the surface of the test specimens was performed using a scanning electron microscope (SEM) (Sigma VP, Carl Zeiss AG, Oberkochen, Germany) at two time points. Time point T0 shows the three processed test specimen groups after modification by INP in Greifswald and the control group without HFG-1. Time point T1 shows the four different surfaces after the 24 h cultivation of HGF-1 on the test specimens.

The test specimens with adhered cells had to be fixed in advance. For fixation, 18 mL polyvinylpyrrolidone/sodium nitrite was mixed in sodium cacodylate buffer and 2 mL glutardialdehyde and incubated for 1 h at 4 °C. The test samples were then rinsed with 0.1 M sodium cacodylate buffer (pH 7.3) three times for 10 min each. For glycocalyx preparation, the test specimens were stored in arginine HCl solution at room temperature for 18 h and then rinsed with distilled water three times for 5 min each. Then, the specimens were placed in a solution of tannin/guanidine HCl solution at room temperature for 5.5 h each. This was followed by rinsing for 5 min with distilled water and three times for 5 min each with sodium cacodylate buffer. Osmylation was performed by incubating the test specimens in 1% OsO_4 in sodium cacodylate buffer for 30 min and then placing them in sodium cacodylate buffer three times for 10 min each. For dehydration or drying, the test specimens remained in 50%, 70%, 90%, and absolute isopropanol for 15 min each and then in 50%, 75%, and 100% acetone (with isopropanol) for 15 min each. After transferring the test specimens to acetone, they were coated with gold–palladium in the SDC 050 sputter coater (BAL-TEC AG Negruet 7, FL-9496, Balzers, Liechtenstein).

Using the computer software SmartSEM (Carl Zeiss AG, Oberkochen, Germany), standardized imaging was performed in a variable low vacuum, with a voltage of 2–10 kV and a magnification of 500×.

3. Results

After the coating procedures, the specimen surfaces had varying surface textures which were analyzed before the cell culture experiments took place.

XPS measurements showed the chemical composition of the coated surfaces. In the case of the hydroxyapatite (HAp), a chemically homogeneous layer was measured: ~50 at% O, ~15 at% C, ~15 at% Ca, ~10 at% P, <5 at% Na, and traces of Si, Mg, and F. Zr is visible from the substrate at the edge, which can be attributed to shading by the sample holder (Figure 1A).

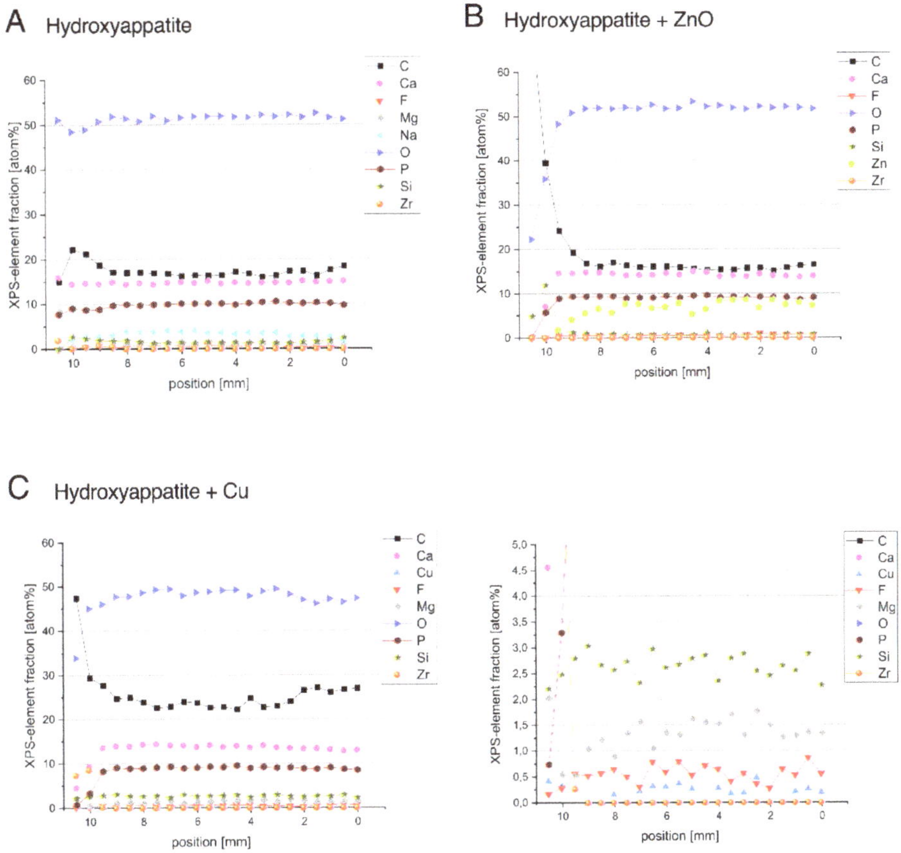

Figure 1. XPS analyses show the chemical composition of specimens after coating with hydroxyapatite (**A**) and hydroxyapatite with zinc (**B**) or copper (**C**). On the right side, an additional extract is shown for better visualization of the amount of copper.

The specimens coated with HAp in combination with 3 wt% ZnO resulted in ~50 at% O, ~15 at% C, ~15 at% Ca, ~10 at% P, and ~9–5 at% Zn, with the Zn content becoming slightly lower towards the left edge. There were also detectable traces of F and Si. Deviations from homogeneity were edge effects caused by the specimen holder during the coating process (Figure 1B).

Coatings with HAp in combination with 3 wt% Cu showed a chemical composition of ~50 at% O, ~25 at% C, ~15 at% Ca, and ~10 at% P. Cu could also be detected, but was measured at only ~0.3 at%; the reason for this is that the Cu was masked by the HAp and the information depth of the XPS analysis is only about 10 nm. Due to the signal-to-noise ratio, it is not possible to say clearly whether this was oxidized or metallic copper. Furthermore, traces of Si, Mg, and F were found (Figure 1C).

Upon visually examining the uncoated surfaces, they appeared smooth and absent of visible irregularities, whereas the coated specimens were characterized by a white smooth surface (Figure 2A). SEM images at 500× magnification confirmed this observation. The untreated specimens of the control group showed a smooth surface with very fine scratches due to the machining process. In contrast, test specimens with hydroxyapatite coating showed an uneven cloud-like surface (Figure 2B).

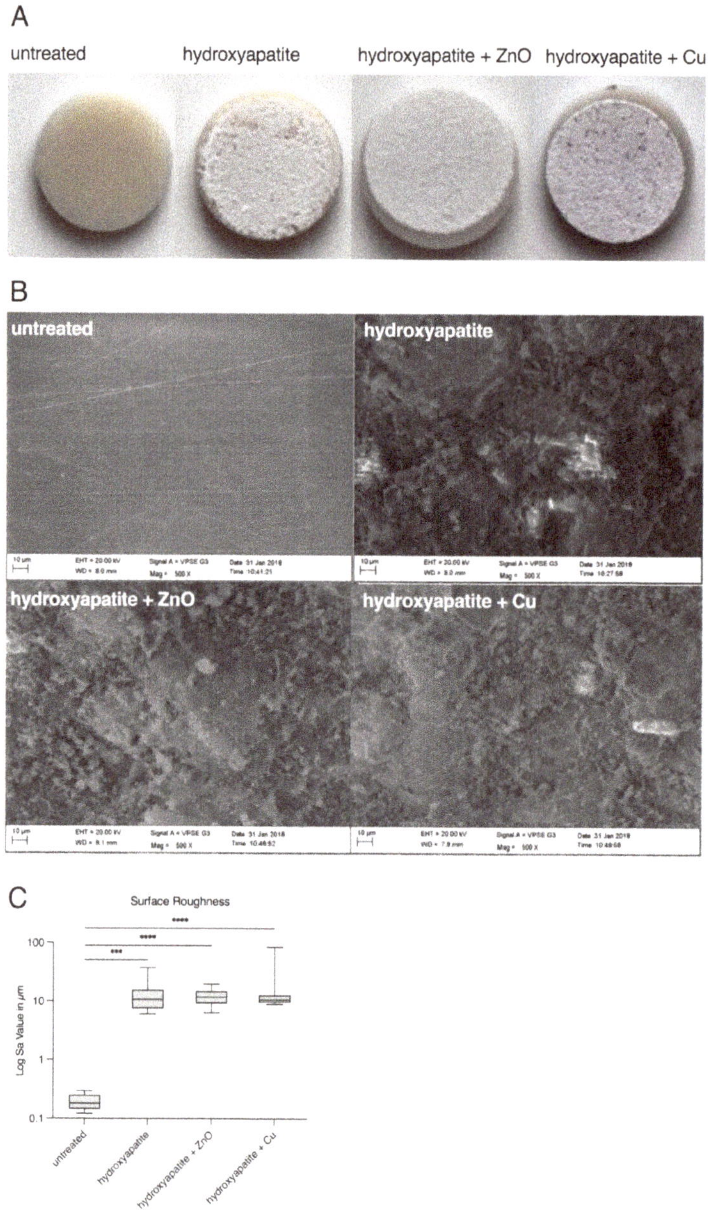

Figure 2. Morphology of specimen surfaces before cell cultivation. (**A**) Photos of specimens. From left to right: untreated, coated with hydroxyapatite, coated with hydroxyapatite + ZnO, and coated with hydroxyapatite + Cu. (**B**) SEM images of specimen surfaces (500× magnification). UL: untreated; UR: coated with hydroxyapatite; LL: coated with hydroxyapatite and zinc oxide (ZnO); LR: coated with hydroxyapatite and copper (Cu). (**C**) Surface roughness (Log Sa values in µm) of untreated and plasma-spayed surfaces. Statistically relevant differences ($p < 0.05$) were found between untreated and treated specimens. *** $p < 0.0005$; **** $p < 0.00005$.

The surfaces were also examined quantitatively. The surface texture (measured in Sa) increased significantly for the hydroxyapatite coatings (Figure 2C, Table 2). The median surface texture was 0.1804 µm for the untreated zirconia specimens. The values for pure hydroxyapatite resulted in a median value of 10.55 µm, hydroxyapatite with zinc oxide was 11.75 µm and hydroxyapatite with copper was 10.47 µm.

Table 2. Descriptive analysis of surface roughness presented in Figure 2C. Median, percentiles, and minimum and maximum Sa values, measured in µm, are shown.

	n =	Median	25% Percentile	75% Percentile	Minimum	Maximum
untreated	12	0.18	0.14	0.24	0.12	0.29
hydroxyapatite	12	10.55	7.41	15.61	5.99	37.30
hydroxyapatite + ZnO	12	11.75	9.08	14.95	6.26	19.68
hydroxyapatite + Cu	12	10.47	9.47	12.80	8.88	84.59

After the detailed characterization of the chemical composition and surface properties, the response of human gingival fibroblasts was investigated. Green value determination after 24 h and 72 h did not yield consistent results for all materials. A positive attachment of HGF-1 cells onto uncoated specimens in terms of number and spreading was observed after 24 h (Figure 3A,B and Table 3). Notably, more HGF-1 cells were observed after 72 h, indicating that the adhered cells were viable and able to proliferate (Figure 3A–C, Tables 3 and 4). In contrast, markedly fewer cells adhered to specimens with hydroxyapatite, hydroxyapatite + ZnO, and hydroxyapatite + Cu (Figure 3A,C and Table 4). However, despite a lower attachment of HGF-1 cells to hydroxyapatite and hydroxyapatite + ZnO, a slightly increased green fluorescence was determined after 72 h, indicating that HGF-1 cells were viable and able to proliferate (Figure 3A,C and Table 4). In contrast, markedly fewer HGF-1 cells were found on hydroxyapatite + Cu specimens after 72 h as compared to 24 h (Figure 3C, Table 4), which might be attributed to the decreased adherence of the cells. Indeed, confocal laser scanning microscopy data revealed a more round-shaped morphology of HGF-1 cells on hydroxyapatite + Cu specimens in comparison to a more elongated and fibroblast morphology of cells on untreated, hydroxyapatite-coated, and hydroxyapatite + ZnO-coated specimens (Figure 3A).

Table 3. Descriptive analysis (median, percentiles, minimum, and maximum) of HGF-1 green value after 24 h of cell cultivation measured in arbitrary units.

	n =	Median	25% Percentile	75% Percentile	Minimum	Maximum
untreated	12	2.99	1.86	3.51	1.076	4.318
hydroxyapatit	12	1.08	0.83	1.82	0.6126	2.277
hydroxyapatite + ZnO	12	1.17	0.90	1.80	0.80	2.15
hydroxyapatite + Cu	12	0.67	0.49	1.31	0.08	1.55

A more in-depth analysis of the morphology of HGF-1 cells seeded onto uncoated and coated specimens was achieved using scanning electron microscopy (Figure 4). Briefly, HGF-1 cells exhibited a typical flat and elongated fibroblast morphology on uncoated specimens. The 500× magnification showed thin, small filamentous secretions, suggestive of proteins and actin filaments (Figure 4).

Table 4. Descriptive analysis (median, percentiles, minimum, and maximum) of HGF-1 green value after 24 h and 72 h of cell cultivation measured in arbitrary units. Significant differences in the results after 24 h and 72 h are highlighted in grey.

	Cell Cluture	$n =$	Median	25% Percentile	75% Percentile	Minimum	Maximum	Significance $p \leq 0.05$
untreated	24 h	6	4.06	3.63	4.76	2.73	5.07	<0.0001
	72 h	6	5.56	5.15	6.73	4.99	6.87	
hydroxyapatite	24 h	6	1.97	1.45	2.32	1.40	2.43	0.2841
	72 h	6	2.42	2.24	2.77	2.17	3.01	
hydroxyapatite + ZnO	24 h	6	1.97	1.59	2.33	1.26	2.81	0.4663
	72 h	6	2.24	1.95	3.24	1.66	3.33	
hydroxyapatite + Cu	24 h	6	1.33	0.89	1.60	0.85	1.83	0.7891
	72 h	6	0.97	0.77	1.14	0.71	1.30	

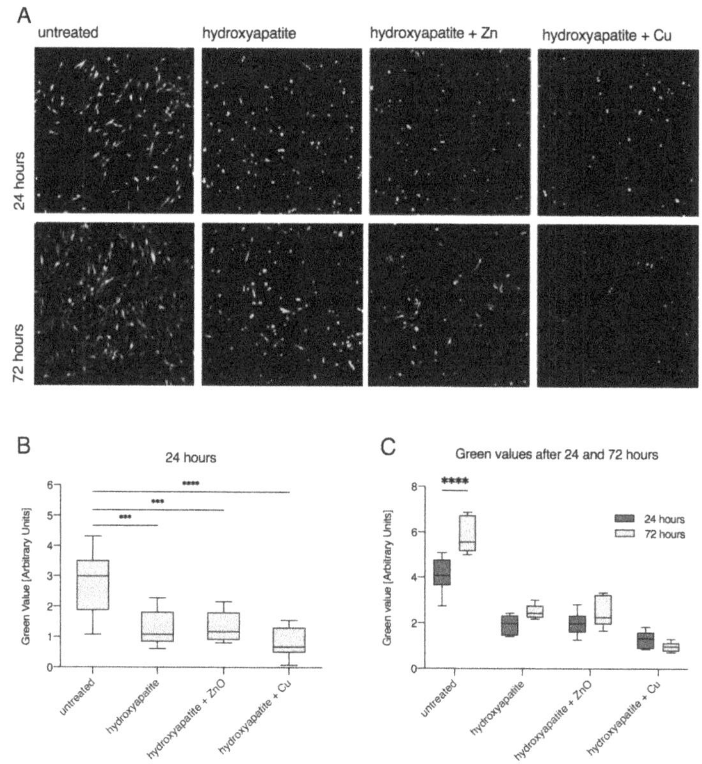

Figure 3. Adhesion and proliferation of human gingival fibroblasts (HGF-1) on uncoated and coated zirconia surfaces. (**A**) Confocal laser scanning microscopy images of specimens. HGF-1 is shown in green (green value) by the fluorescent dye CMFDA. Shown are representative images after 24 h and 72 h of cell cultivation. (**B**) HGF-1 green value after 24 h of cell cultivation. Statistically relevant differences ($p < 0.05$) were found between untreated and treated specimens. *** $p < 0.0005$; **** $p < 0.00005$. (**C**) Upon comparing results for the HGF-1 green value after 24 h and 72 h of cultivation, statistically significant increases ($p < 0.05$) were found for untreated specimens. **** $p < 0.00005$.

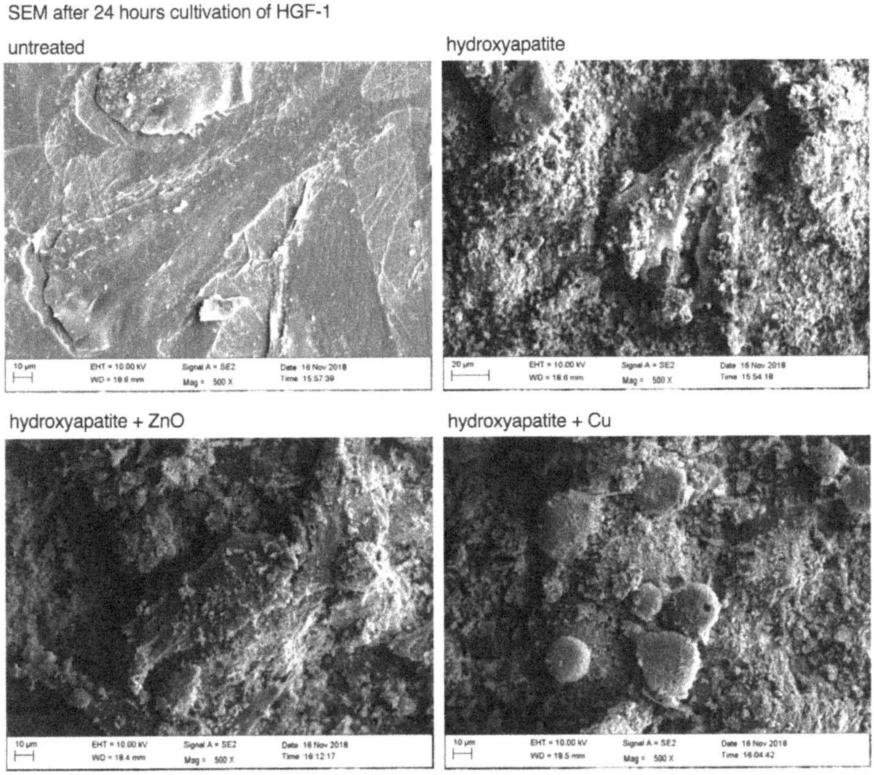

Figure 4. Representative SEM images of the specimens with attached HGF-1 cells at 500× magnification. UL: untreated; UR: coated with hydroxyapatite; LL: coated with hydroxyapatite +ZnO; LR: coated with hydroxyapatite + Cu.

In contrast, HGF-1 cells on coated surfaces exhibited a markedly different morphology. On surfaces coated with hydroxyapatite and hydroxyapatite +ZnO, HGF-1 cells appeared slightly spread out and had conspicuous deposits on the cell surface, whereas cells on surfaces coated with hydroxyapatite + Cu were markedly smaller and, more importantly, spherical (Figure 4). The SEM data suggested a porous membrane of HGF-1 cells on hydroxyapatite + Cu concomitant with barely visible filamentous protein structures. Briefly, these findings were in line with the confocal laser scanning microscopy data (Figure 3A) and likely indicated a reduced capability of HGF-1 cells to adhere to hydroxyapatite + Cu-coated specimens.

4. Discussion

Until now, titanium was considered the most important material for use in dental implants. Its high load-bearing capacity and good compatibility with bone and soft tissue makes titanium an almost perfect material. Although failures do occur, a median 10-year implant survival rate of 94.6% has been reported [19]. One factor leading to failure of dental implants could be the corrosion of titanium, which cannot be prevented in metals [20]. So far, the main alternatives to titanium are made with zirconium dioxide. Some of these are already used as dental implants, but also as abutments. Nevertheless, there is still potential for further developments in preclinic.

For any in vitro model, there are limitations that must be considered. In this model, only one aspect of the complex process between the abutment surface and the biosystem was investigated. The growth and adhesion behavior in association with other mucosal cells, the implant environment, the loading conditions at the tissue-implant interface, the surgical procedure for implant application, and a prolonged healing time were not considered. These aspects can only be studied in a living organism. However, the simplicity of the model used here enables a reproducible environment, a standardized investigation of further cell species from the oral mucosa, and accordingly comparable results.

In addition to the material itself, the characteristics of the surface play a major role in cellular attachment. Cells are known to behave differently on different surface morphologies, and a rough surface favors bacterial attachment [21–23]. On polished surfaces, human gingival fibroblasts (HGF) spread without any particular orientation [24]. On the other hand, fine micro-textures, such as grooves or clusters, may favor the cell configuration, orientation, and proliferation of HGF [24–26]. Thus, the surface of these materials must be characterized in addition to the proliferation itself. It is also known that surface roughness up to 0.34 μm Sa does not prevent the attachment of HGF [27]. This could explain the poorer proliferation and adhesion in addition to the effect of hydroxyapatite, zinc oxide, or copper on HGF cells. Nevertheless, slightly increased proliferation was observed in all groups except hydroxyapatite + copper after 72 h of cell cultivation. In our study, roughness values in the range of 10 μm were observed (Figure 2C, Table 2). In plasma coating, surface roughness can be influenced by the choice of powder and spray parameters [28], and thermal post-treatment to reduce roughness is also possible. The modification of the surface texture should be considered in further studies for a possible increase in cell proliferation. Furthermore, it has been shown that an increase in the roughness of implant parts with a connection to soft tissue can also lead to an increase in the inflammation of peri-implant tissue [29]. Although it cannot be excluded that antibacterial substances would be antagonistic, this risk must be taken into consideration when developing new abutments.

In the future, if hydroxyapatite surfaces can be made more attractive to fibroblasts and other cells, it is also reasonable to perform a deeper investigation of adhesion and proliferation. Changes at the DNA and protein level related to roughness have already been shown in osteoblasts [30].

In the present study, antimicrobial essential trace elements of zinc oxide and copper were used. The properties of zinc oxide inhibit the acid production of bacteria in the oral cavity [31], and inhibition has been observed in Gram-negative bacteria [32]. The trace element copper shows antibacterial effects on *S. mutans* and *P. gingivalis* in vitro [33]. However, selective inhibition may result in the greater growth of other species [34]. Abrahamsson et al. showed that the ideal transmucosal surface of an implant should promote the rapid growth of soft tissue cells while reducing bacteria [35]. However, antibacterial substances should not also have a negative influence on the cell growth of fibroblasts. A shown here, HGF-1 cells seeded onto hydroxyapatite + Cu possessed a more spherical morphology, suggesting that the ability of the cells to adhere was hampered, which was further associated with impaired proliferation. Whether these findings indicate that antibacterial compounds such as Cu could have a negative effect on the adhesion and proliferation of gingival fibroblasts remains to be elucidated in further studies.

5. Conclusions

Despite the limitations caused by the in vitro model, some findings were obtained by using non-immortalized human fibroblasts (HGF-1) in this study. The results show that HGF-1 cells adhere and proliferate well on machined zirconia, whereas additional plasma coatings with hydroxyapatite or hydroxyapatite mixtures with zinc oxide and copper do not result in increased adhesion or proliferation. Even the machined and untreated zirconia surface appears to be sufficiently attractive to fibroblasts.

Author Contributions: O.B. wrote the original draft, supervised, and conceived parts of the project. M.-C.S. performed the majority of the experiments and reviewed the literature. K.B. wrote parts of the original draft. H.T. and K.F. supervised and performed the plasma atmospheric coatings. A.Q. conducted the XPS analysis. E.A.N. and W.H.A. conceived parts of the data validation and supervised the SEM analysis. A.P. and T.D. supervised the project and corrected the manuscript. All authors have read and agreed to the published version of the manuscript.

Funding: This project was financially supported by the Fördergemeinschaft Zahnmedizin e.V. (Witten, Germany). Vita Zahnfabrik (Bad Säckingen, Germany) provided us with the test specimens. Neither the association nor the company had any influence on the design or execution of this study.

Institutional Review Board Statement: Not applicable.

Informed Consent Statement: Not applicable.

Data Availability Statement: The data from this study can be shared by the corresponding author upon request.

Acknowledgments: We thank Martin Polak (INP Greifswald, Greifswald, Germany) for supporting the surface coating. We also thank Silvia Keil (Witten/Herdecke University, Witten, Germany) for helping with cell culture experiments and Susanne Haußmann and Katja Fröhlich (Witten/Herdecke University, Witten, Germany) for preparing the SEM images.

Conflicts of Interest: The authors declare no conflict of interest.

References

1. Brånemark, P.-I.; Breine, U.; Adell, R.; Hansson, B.; Lindström, J.; Ohlsson, Å. Intra-osseous anchorage of dental prostheses: I. Experimental studies. *Scand. J. Plast. Reconstr. Surg.* **1969**, *3*, 81–100. [CrossRef] [PubMed]
2. Mombelli, A.; Van Oosten, M.; Schürch, E., Jr.; Lang, N. The microbiota associated with successful or failing osseointegrated titanium implants. *Oral Microbiol. Immunol.* **1987**, *2*, 145–151. [CrossRef] [PubMed]
3. Jepsen, S.; Berglundh, T.; Genco, R.; Aass, A.M.; Demirel, K.; Derks, J.; Figuero, E.; Giovannoli, J.L.; Goldstein, M.; Lambert, F.; et al. Primary prevention of peri-implantitis: Managing peri-implant mucositis. *J. Clin. Peiodontol.* **2015**, *42*, S152–S157. [CrossRef] [PubMed]
4. Renvert, S.; Polyzois, I. Risk indicators for peri-implant mucositis: A systematic literature review. *J. Clin. Periodontol.* **2015**, *42*, S172–S186. [CrossRef]
5. Lang, N.P.; Bosshardt, D.D.; Lulic, M. Do mucositis lesions around implants differ from gingivitis lesions around teeth? *J. Clin. Periodontol.* **2011**, *38*, 182–187. [CrossRef]
6. Al Mugeiren, O.M.; Baseer, M.A. Dental implant bioactive surface modifiers: An update. *J. Int. Soc. Prev. Community Dent.* **2019**, *9*, 1. [CrossRef]
7. Sculean, A.; Chappuis, V.; Cosgarea, R. Coverage of mucosal recessions at dental implants. *Periodontol. 2000* **2017**, *73*, 134–140. [CrossRef]
8. Ivanovski, S.; Lee, R. Comparison of peri-implant and periodontal marginal soft tissues in health and disease. *Periodontol. 2000* **2018**, *76*, 116–130. [CrossRef]
9. Moon, I.S.; Berglundh, T.; Abrahamsson, I.; Linder, E.; Lindhe, J. The barrier between the keratinized mucosa and the dental implant: An experimental study in the dog. *J. Clin. Periodontol.* **1999**, *26*, 658–663. [CrossRef]
10. Wang, Y.; Zhang, Y.; Miron, R.J. Health, maintenance, and recovery of soft tissues around implants. *Clin. Implant Dent. Relat. Res.* **2016**, *18*, 618–634. [CrossRef]
11. Barillas, L.; Testrich, H.; Cubero-Sesin, J.M.; Quade, A.; Vargas, V.I.; Polák, M.; Fricke, K. Bioactive plasma sprayed coatings on polymer substrates suitable for orthopedic applications: A study with PEEK. *IEEE Trans. Radiat. Plasma Med. Sci.* **2018**, *2*, 520–525. [CrossRef]
12. Fukumoto, M.; Nie, M.; Yasui, T. Preparation and evaluation of ordinary attritor milled Ti-Al powders and corresponding thermal sprayed coatings. *Mater. Trans.* **2006**, *47*, 1717–1722. [CrossRef]
13. Sun, L.; Berndt, C.C.; Gross, K.A.; Kucuk, A. Material fundamentals and clinical performance of plasma-sprayed hydroxyapatite coatings: A review. *J. Biomed. Mater. Res.* **2001**, *58*, 570–592. [CrossRef] [PubMed]
14. Auclair-Daigle, C.; Bureau, M.; Legoux, J.G.; Yahia, L.H. Bioactive hydroxyapatite coatings on polymer composites for orthopedic implants. *J. Biomed. Mater. Res. Part A* **2005**, *73*, 398–408. [CrossRef]
15. Beauvais, S.; Decaux, O. Plasma sprayed biocompatible coatings on PEEK implants. In Proceedings of the 2007 International Thermal Spray Conference, Beijing, China, 14–16 May 2007; pp. 371–377.
16. Benčina, M.; Resnik, M.; Starič, P.; Junkar, I. Use of plasma technologies for antibacterial surface properties of metals. *Molecules* **2021**, *26*, 1418. [CrossRef]
17. Jung, R.E.; Grohmann, P.; Sailer, I.; Steinhart, Y.N.; Fehér, A.; Hämmerle, C.; Strub, J.R.; Kohal, R. Evaluation of a one-piece ceramic implant used for single-tooth replacement and three-unit fixed partial dentures: A prospective cohort clinical trial. *Clin. Oral Implant. Res.* **2016**, *27*, 751–761. [CrossRef]
18. Ruan, H.; Fan, C.; Zheng, X.; Zhang, Y.; Chen, Y. In vitro antibacterial and osteogenic properties of plasma sprayed silver-containing hydroxyapatite coating. *Chin. Sci. Bull.* **2009**, *54*, 4438–4445. [CrossRef]

19. Moraschini, V.; Poubel, L.D.C.; Ferreira, V.; Barboza, E.D.S. Evaluation of survival and success rates of dental implants reported in longitudinal studies with a follow-up period of at least 10 years: A systematic review. *Int. J. Oral Maxillofac. Surg.* **2015**, *44*, 377–388. [CrossRef]
20. Noumbissi, S.; Scarano, A.; Gupta, S. A Literature review study on atomic ions dissolution of titanium and its alloys in implant dentistry. *Materials* **2019**, *12*, 368. [CrossRef]
21. Quirynen, M.; Bollen, C. The influence of surface roughness and surface-free energy on supra-and subgingival plaque formation in man: A review of the literature. *J. Clin. Periodontol.* **1995**, *22*, 1–14. [CrossRef]
22. Rimondini, L.; Cerroni, L.; Carrassi, A.; Torricelli, P. Bacterial colonization of zirconia ceramic surfaces: An in vitro and in vivo study. *Int. J. Oral Maxillofac. Implant.* **2002**, *17*, 739–798.
23. Siegrist, B.E.; Brecx, M.C.; Gusberti, F.A.; Joss, A.; Lang, N.P. In vivo early human dental plaque formation on different supporting substances. A scanning electron microscopic and bacteriological study. *Clin. Oral Implant. Res.* **1991**, *2*, 38–46. [CrossRef] [PubMed]
24. Köunönen, M.; Hormia, M.; Kivilahti, J.; Hautaniemi, J.; Thesleff, I. Effect of surface processing on the attachment, orientation, and proliferation of human gingival fibroblasts on titanium. *J. Biomed. Mater. Res.* **1992**, *26*, 1325–1341. [CrossRef]
25. Mustafa, K.; Lopez, B.S.; Hultenby, K.; Wennerberg, A.; Arvidson, K. Attachment and proliferation of human oral fibroblasts to titanium surfaces blasted with TiO2 particles. A scanning electron microscopic and histomorphometric analysis. *Clin. Oral Implant. Res.* **1998**, *9*, 195–207. [CrossRef] [PubMed]
26. Pae, A.; Lee, H.; Kim, H.-S.; Kwon, Y.-D.; Woo, Y.-H. Attachment and growth behaviour of human gingival fibroblasts on titanium and zirconia ceramic surfaces. *Biomed. Mater.* **2009**, *4*, 025005. [CrossRef]
27. Mustafa, K.; Odén, A.; Wennerberg, A.; Hultenby, K.; Arvidson, K. The influence of surface topography of ceramic abutments on the attachment and proliferation of human oral fibroblasts. *Biomaterials* **2005**, *26*, 373–381. [CrossRef]
28. Tsui, Y.; Doyle, C.; Clyne, T. Plasma sprayed hydroxyapatite coatings on titanium substrates Part 1: Mechanical properties and residual stress levels. *Biomaterials* **1998**, *19*, 2015–2029. [CrossRef]
29. Degidi, M.; Artese, L.; Piattelli, A.; Scarano, A.; Shibli, J.A.; Piccirilli, M.; Perrotti, V.; Iezzi, G. Histological and immunohistochemical evaluation of the peri-implant soft tissues around machined and acid-etched titanium healing abutments: A prospective randomised study. *Clin. Oral Investig.* **2012**, *16*, 857–866. [CrossRef]
30. Brett, P.; Harle, J.; Salih, V.; Mihoc, R.; Olsen, I.; Jones, F.; Tonetti, M. Roughness response genes in osteoblasts. *Bone* **2004**, *35*, 124–133. [CrossRef]
31. Phan, T.-N.; Buckner, T.; Sheng, J.; Baldeck, J.; Marquis, R. Physiologic actions of zinc related to inhibition of acid and alkali production by oral streptococci in suspensions and biofilms. *Oral Microbiol. Immunol.* **2004**, *19*, 31–38. [CrossRef]
32. Fröber, K.; Bergs, C.; Pich, A.; Conrads, G. Biofunctionalized zinc peroxide nanoparticles inhibit peri-implantitis associated anaerobes and *Aggregatibacter actinomycetemcomitans* pH-dependent. *Anaerobe* **2020**, *62*, 102153. [CrossRef] [PubMed]
33. Liu, R.; Memarzadeh, K.; Chang, B.; Zhang, Y.; Ma, Z.; Allaker, R.P.; Ren, L.; Yang, K. Antibacterial effect of copper-bearing titanium alloy (Ti-Cu) against Streptococcus mutans and *Porphyromonas gingivalis*. *Sci. Rep.* **2016**, *6*, 29985. [CrossRef] [PubMed]
34. Greenstein, G.; Polson, A. The role of local drug delivery in the management of periodontal diseases: A comprehensive review. *J. Periodontol.* **1998**, *69*, 507–520. [CrossRef] [PubMed]
35. Abrahamsson, I.; Berglundh, T.; Wennström, J.; Lindhe, J. The peri-implant hard and soft tissues at different implant systems. A comparative study in the dog. *Clin. Oral Implant. Res.* **1996**, *7*, 212–219. [CrossRef]

materials

MDPI

Article

Surface Characteristics of High Translucent Multilayered Dental Zirconia Related to Aging

Flavia Roxana Toma [1], Mihaela Ionela Bîrdeanu [2], Ion-Dragoș Uțu [3], Roxana Diana Vasiliu [1], Lavinia Cristina Moleriu [4] and Liliana Porojan [1,*]

[1] Department of Dental Prostheses Technology (Dental Technology), Center for Advanced Technologies in Dental Prosthodontics, Faculty of Dental Medicine, "Victor Babeș" University of Medicine and Pharmacy Timișoara, EftimieMurgu Sq. No. 2, 300041 Timisoara, Romania; flavia.toma@umft.ro (F.R.T.); roxana.vasiliu@umft.ro (R.D.V.)

[2] National Institute for Research and Development in Electrochemistry and Condensed Matter, 300569 Timisoara, Romania; mihaelabirdeanu@gmail.com

[3] Department of Materials and Fabrication Engineering, Politehnica University Timişoara, Bd. Mihai Viteazu nr.1, 300222 Timisoara, Romania; dragos.utu@upt.ro

[4] Department of Functional Science, "Victor Babeș" University of Medicine and Pharmacy Timișoara, EftimieMurgu Sq. No. 2, 300041 Timisoara, Romania; moleriu.lavinia@umft.ro

* Correspondence: sliliana@umft.ro

Abstract: (1) Background: The purpose of this study was to evaluate the differences in terms of surface characteristics (roughness, topography, microhardness) among layers for multi-layered high translucent and super-high translucent zirconia and the influence of finishing and aging on surface characteristics and microstructure. (2) Methods: Three types of translucent multilayer zirconia were evaluated: STML (4Y-TZP); IPS e.maxZirCAD CEREC/in Lab MT Multi (4Y-TZP + 5Y-TZP); CeramillZolidfx ML (5Y-TZP). Ninety-six plate-shaped samples (32 for a material), 16 mm × 14 mm × 1 mm size, were cut with a precision disc, polished on both sides with sand papers and sintered respecting the manufacturer's protocol. Half of the specimens (16) were finished by polishing and the other half by glazing and then equally divided into one control group and one group subject to aging by autoclaving (1 h, 134 °C, 0.2 MPa), resulting in four groups of eight samples, for each zirconia. The specimens were evaluated in three areas: cervical, medium, incisal-of each glazed or polished surface, before and after aging. Tests were performed to determine the surface roughness using a profilometer; the surface topography by an atomic force microscope (AFM) and a scanning electron microscope (SEM). Microhardness was recorded using a microhardness tester. Statistical analyses were performed using two-way ANOVA test, unpaired sample *t*-Test, paired sample *t*-Test ($\alpha = 0.05$) and Pearson's correlation. (3) Results: Before and after autoclaving, for glazed samples significance ($p < 0.05$) higher surface roughness, respectively lower microhardness in comparison with the polished group was assessed. No significant differences ($p > 0.05$) were reported between the three areas, on glazed or polished surfaces of a material. Although, after aging an increase in surface roughness was observed both on glazed and polished samples, statistical differences were found for STML ($p < 0.05$). No significant differences ($p > 0.05$) concerning microhardness among the same areas, on glazed and polished surfaces, recorded before and after aging, except CeramillZolidfx ML glazed samples. (4) Conclusions: For tested zirconia materials no significant differences among layers were registered regarding surface characteristics. Surface treatment (glazing or polishing) has a significant impact on surface roughness and microhardness. Both before and after aging, the surface roughness values for the glazed samples were higher than for those polished. The super translucent 4Y-TZP material was more affected by aging compared to the super-high translucent 5Y-TZP material. The combined material revealed similarities for each layer corresponding to the microstructure.

Keywords: multilayer zirconia; microroughness; surface topography; microhardness; aging

Citation: Toma, F.R.; Bîrdeanu, M.I.; Uțu, I.-D.; Vasiliu, R.D.; Moleriu, L.C.; Porojan, L. Surface Characteristics of High Translucent Multilayered Dental Zirconia Related to Aging. *Materials* **2022**, *15*, 3606. https://doi.org/10.3390/ma15103606

Academic Editors: John G. Fisher and Sangwon Park

Received: 11 April 2022
Accepted: 16 May 2022
Published: 18 May 2022

1. Introduction

To eliminate disadvantages as opacity, risk of chipping, substantial reduction of dental structure and also due aesthetics, biocompatibility and chemical stability compared to framework or conventional metal–ceramic, monolithic zirconia restorations are widely used in dentistry [1]. In order to obtain materials with high translucency, but also resistant, new grades of zirconia have been developed, pre-sintered blocks being available as pre-shaded in various tonalities or multilayered with gradients of chroma [2]. The translucent dental zirconia generations involved modifications of the crystal structure, increasing the content of Y_2O_3 (Yttrium) and cubic phase, resulting in two polycrystalline materials: 5Y-TZP and 4Y-TZP (partially stabilized zirconia with 5 mol% and respectively 4 mol% Yttrium) [3,4]. The main phase of zirconia is decisive for the individual characteristics of each zirconia system [5,6].

These compositional differences can influence three important clinical properties of the material: translucency, toughness, and resistance to hydrothermal aging [7].

The strategy for improving the optical performance of material imply microstructural variations—the higher Yttrium content influences the coefficient of thermal expansion, the size of grains (they are larger), reduce the residual porosity, resulting fewer boundaries, less birefringence, and scattering of the light [8], leading to higher translucency [6]. A recent study related that 5Yis 33%, and 4Yis 13% more translucent than 3Y-zirconia (3 mol% Yttrium stabilized-tetragonal zirconia polycrystal) [9]. The nanoparticles zirconia is presumed to be the next progress in evolution of this material [10–12].

In the microstructure, when stress occurs, the tetragonal phase ensures the ability to withstand compression, stopping the progression of the micro-cracks and elimination of microstructure defects [13], inducing transformation toughening. This mechanism is reduced for zirconia with higher cubic content. The super translucent 4Y-TZP (fourth-generation) contains 4 mol% Y_2O_3 and around 25% cubic phase. For super-high translucent 5Y-TZP (third-generation), the Y_2O_3 content was increased to 5 mol% and the quantity of the cubic phase rises up to 50%, also the amount of tetragonal phase decreases, leading to a lower possibility of tetragonal to monoclinic phase transformation, with reduction in strength and fracture toughness [8]; that is a disadvantage of these new translucent materials and the clinicians should use the 4Y and 5Y zirconia when the aesthetic needs are required.

Autoclave treatment induce an accelerate aging of zirconia [14,15].

Negative effects were reported after simulated accelerating aging of zirconia—LTD (low temperature degradation), by autoclaving [16], but limited types of materials were evaluated in short-term aging, hence different studies on multilayered translucent zirconia are essential for clinical applications. The aging process is associated with a slow tetragonal to monoclinic phase transformation, 4% volume increase of particles, micro-cracks with water penetration, grain detachment, decrease in hardness and fracture strength, surface roughening and degradation at the surface (and subsurface) which is in contact with moisture [1,17,18]. It was observed that hydrothermal and mechanical impact could cause unexpected phase transformation [19]. However, cubic zirconia undergoes almost no monoclinic transformation and hydrothermal degradation. Many studies revealed that after aging, 5Y-TZP was less affected than 3Y-TZP, but both experienced a strength reduction [20].

It was observed that the surface roughness and the microstructure characteristics can affect the strength of the material [21] and be relevant factors that influence the appearance, the final color, or bacterial accumulation [22,23].

The monolithic multilayer zirconia has polychromatic, pre-colored layers—enamel (incisal), transition 1 and 2 (medium), dentin-body (cervical)—simulating the shade and translucency of the natural teeth; with chroma and opacity increasing toward gingival area; the enamel layer looking the most transparent, while the dentin layer appears more opaque owingto the considerable low values for light permeability.

Several studies have reported that the first introduced multilayer zirconia (Katana, Kuraray Noritake), presented the same content of yttrium, respectively cubic phase in the layers of the material; among these being only the color difference given by the pigment composition [24]. Later, multilayer zirconia with different composition and microstructure in the same material (IPS e.maxZirCad Multi, IvoclarVivadent) was introduced.

The layers result as various formulations depending on the quantity of stabilizer and chemical composition. Because of these values, potential differences in the physical properties in the individual layers occur [6,25,26].

However, there are contradictory reports regarding the negative effects of the incorporation of oxide pigments on strength of zirconia [27,28] and by interface between layers was evidenced 30% lower strength than the bulk of material [29].

To improve aesthetics, polishing and glazing as surface treatments are applied [23,30]. It has often been shown that the polishing ability depends on microstructure of the material [31]. The slight polishing causes transformation nucleation usually around scratches due to tensile residual stress [32] and an increase in the monoclinic phase, may strengthen the material, but implies a directly proportional increase in surface roughness [33,34]. The gentle grinding and polishing using a small-grit (15–30μm) tool and a low-speed handpiece, cause less-extensive tetragonal to monoclinic transformation and induce minor or no obvious flaws [35–37] in zirconia with different Yttrium concentrations [38].

Glazing as a laboratory procedure is achieved through applying a blend of colorless glass powder and liquid layer that reduced the roughness, seals the pores, and smoothensthe ceramic surface [39,40]. The appearance of the glazed surface can be influenced by the layer deposition technique, the special brush, layer thickness, and uniformity of zirconia interface features, sometimes resulting in an irregular wavy surface with small defects. According to various results, after aging, the glaze layer may be negatively affected, broken, or separated from thematerial [41,42].

It is known, when the dopant concentration is high, the surface roughness is increased (meaning for 8Y-or 5Y-TZP was higher in comparison to 3Y-TZP) [43,44]; also, it causesan increase incubic phase amount and lowers the strength and fracture toughness, [1,6] and lowers the mechanical properties [45]. As a result, the surface treatment as glazing may or not have a protective role; polishing may induce defects and/or phase transformation which can lead to an increase or decrease in strength of the material [32,46–48].

Glazing and polishing enhance the appearance of restorations, decrease the bacterial accumulation, the inflammatory tissue response and the wear of opposing teeth [31–33].

This in vitro study aims to evaluate: (a) The differences in terms of surface characteristics (roughness, topography, micro-hardness) among layers for multi-layered high translucent and super-high translucent zirconia; (b) how glazing and polishing affect the surface properties of the mentioned materials; (c) the influence of LTD (low temperature degradation) as aging method, on surface characteristics and microstructure.

The null hypotheses were: (1) There are not significant differences regarding the surface texture and microhardness in the three areas (cervical, medium, incisal) of the materials; (2) glazing and polishing affect the surface properties of the material; (3) the roughness value increases and microhardness decreases after the aging procedure.

2. Materials and Methods

Three types of commercially multilayered dental zirconia with various translucencies and Yttrium content wasinvestigated in this study: super-translucent zirconia STML (Katana, Kuraray Noritake Dental, Tokio, Japan)-noted ST; IPS e.maxZirCAD CEREC/in Lab MT Multi (IvoclarVivadent AG, Schaan, Liechtenstein)-noted IP and super-high-translucent zirconia CeramillZolidfx ML (Amman Girrbach, AG, Koblach, Austria)-noted CE. The properties of these materials are presented in Table 1.

Table 1. Properties of the materials taken into the study [49–51].

Restauration Material	Manufacturer	Translucency/ Shade	Yttrium Content	Flexural Strength
STML (ST)	Katana, Kuraray Noritake Dental, Japan	Super translucency A2	4 mol%	750 MPa
IPS e.max Zir CAD (IP)	Ivoclar Vivadent AG, Schaan, Liechtenstein	Super-high translucency + Super translucency A2	4 mol%-incisal 4 + 5 mol%-transition 4 mol%-cervical	850 MPa
Ceramill Zolid fx ML (CE)	Amman Girrbach, AG, Koblach, Austria	Super-high translucency A2/A3	5 mol%	700 MPa

2.1. Specimen Preparation

Ninety-six specimens (32 samples of each zirconia) were cut from presintered blocks of each material, with a precision disc, followed by polishing on both sides using #600, #1500, #2000 sandpapers. Each material was sintered in a ceramic furnace LHT 01/17D (Nabertherm, Lilienthal, Germany) 1550 °C/7h-for STML; DekemaAustromat 674 (DEKEMA Dental-Keramiköfen GmbH, Freilassing, Germany) 1500 °C/9 h 50 min-for IPS e.maxZirCAD CEREC/in Lab MT Multi; Programat S2 (IvoclarVivadent, Schaan, Liechtenstein) 1450 °C/8h-for CeramillZolidfx ML according to a protocol recommended by each manufacturer. Half of the specimens (16 samples) were finished only by polishing and the other half by glazing. The finishing procedure was performed on one surface of the plate. Further, each assortment was divided equally into one control group and one group subject to aging by autoclaving (1 h, 134 °C, 0.2 MPa), as LTD (low temperature degradation). Four groups (n = 8) resulted for each material: control glazed (cg), control polished (cp), autoclaved glazed (ag), autoclaved polished (ap). The glazed (g) and the polished (*p*) surfaces were evaluated into three areas corresponding to the layers: cervical (c), medium (m), incisal (i), before, and after aging.

Glazing was performed for all specimens with two thin layers—IPS Ivocolor Glaze Paste (IvoclarVivadent, Schaan, Liechtenstein) in Vita Vacumat 6000 M furnace (Bad Sackingen Germany), in accordance with the manufacturer's indications. For polishing a low-speed device, polishing brush and diamond paste-Renfert Polish (Renfert, Hilzingen, Germany) was used, following the same technique to obtain a smooth and glossy surface. The final dimensions of plate-shaped samples were 16 mm × 14 mm and 1 mm thickness, andweremeasured with a caliper. For a better understanding a flowchart was created, inwhichthe studied groups are described-Figure 1.

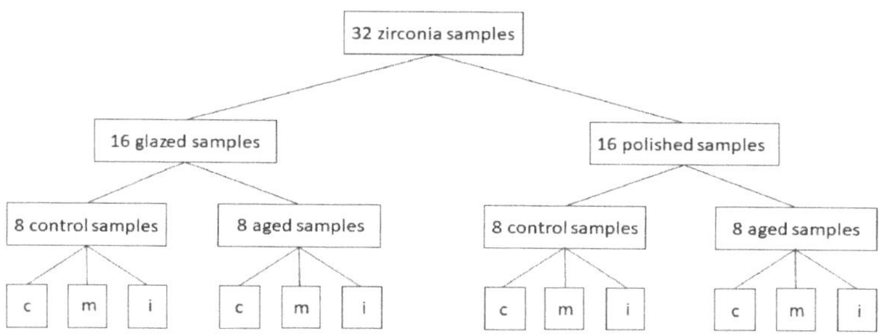

Figure 1. The flowchart of the study in which the main samples and groups were plotted and the way they are divided; c = cervical area, m = medium area, i = incisal area.

2.2. *Hydrothermal Aging*

The samples were subjected to the artificial aging process as LTD (low temperature degradation) using an autoclave (Sterilclave 24 B, Cominox, CarateBrianza, Italy) with distilled water for 1 h, at 134 °C and 0.2 MPa; it is known that 1 h of autoclaving in these conditions would induce on zirconia, the same effects as 3–4 years in vivo [14].

2.3. *Surface Roughness and Topography*

The measurement of surface roughness was made using a contact profilometerSurftest SJ-201(Mitutoyo, Kawasaki, Japan) with a 2 μm diamond stylus. On each level (cervical, medium, incisal), three measurements were recorded in three randomly selected areas, on polished and on the glazed samples, respectively. The values of the Ra (μm) and Rz (μm) parameters were obtained. Ra (μm) is arithmetic average surface roughness and Rz (μm) is maximum surface roughness. The cut-off value was 0.3 mm to a force applied by 0.7 mN.

2.4. *Atomic Force Microscopy (AFM)*

The nanosurface characterization of the specimens was evaluated with an atomic force microscope Nanosurf Easy Scan 2 Advanced Research (Nanosurf AG, Liestal, Switzerland). The values of nanoroughness parameters Sa (nm)-arithmetical mean height and Sq (nm)-root mean square height resulted. For surface topographic analysis, AFM provided three-dimensional images that were achieved with a scan size of 2.2 × 2.2 μm, in non-contact mode.

2.5. *Scanning Electron Microscopy (SEM)*

To investigate the surface modification resulting from different surface treatments, a scanning electron microscope with high resolution field emission—Inspect S (FEI Company, Hillsboro, OR, USA) was used. The assessment of surface characteristics was performed on both samples (glazed, polished), in the three individual colored areas (cervical, medium, incisal).

2.6. *Microhardness Testing*

A microhardness tester (HVS-10A1, Huatec, Bejing, China), was used to assess the surface hardness of glazed, respectively polished surfaces of the samples, on each domain (cervical, medial, incisal) in three randomly selected areas. The indentation technique determined the Vickers Hardness number (VN), by applying a force of 0.3 kg for 10 s.

2.7. *Statistical Analysis*

Two different programs were used for statistical analyses—the IBM SPSS Statistics software (IBM, New York, NY, USA) and JASP (v.16.2, University of Amsterdam, Holland). In the first part was used descriptive statistics, calculating the central tendency and dispersion parameters. To decide upon the statistical type of tests we applied the Shapiro–Wilk test for distribution and obtained that the data are normally distributed (p> 0.05) in most of the cases; for a complete perspective, parametrical and non-parametrical tests were applied. The repeated measures ANOVA test was used for statistical analysis of more than two-time moments/dependent groups. The unpaired t-Test was applied for two different groups (for statistical analysis of variables recorded on glazed and polished sides); the paired t-Test was performed for two-time moments situation (for comparing areas of treated surfaces-before and after aging, for a material and the significant differences because of autoclaving).

A correlation model between Ra and microhardness, was performed—the straightness of the correlation was measured using Pearson's coefficient. For the entire study, a significance level of α =0.05 was established.

3. Results

3.1. Micro-Surface Roughness

Mean values and SD (standard deviation) for Ra (µm), arithmetic average surface roughness and Rz (µm), maximum surface roughness measurements on glazed and polished samples, before aging are summarized in Table 2.

Table 2. Mean values with SD for Ra, Rz parameters of the samples, before aging (n = 8).

		ST cg	ST cp	IP cg	IP cp	CE cg	CE cp
	c	0.064 ± 0.009	0.041 ± 0.009	0.073 ± 0.009	0.043 ± 0.010	0.086 ± 0.010	0.047 ± 0.009
Ra	m	0.065 ± 0.009	0.043 ± 0.010	0.080 ± 0.009	0.044 ± 0.009	0.085 ± 0.010	0.045 ± 0.010
	i	0.067 ± 0.010	0.042 ± 0.010	0.083 ± 0.009	0.046 ± 0.009	0.084 ± 0.008	0.047 ± 0.010
	c	0.445 ± 0.099	0.360 ± 0.094	0.475 ± 0.098	0.366 ± 0.098	0.580 ± 0.096	0.386 ± 0.082
Rz	m	0.466 ± 0.096	0.363 ± 0.091	0.530 ± 0.091	0.371 ± 0.096	0.573 ± 0.098	0.371 ± 0.091
	i	0.493 ± 0.093	0.378 ± 0.100	0.533 ± 0.100	0.385 ± 0.092	0.563 ± 0.090	0.373 ± 0.093

Ra (µm)—arithmetic average surface roughness, Rz (µm)—maximum surface roughness. ST= STML, IP= IPS e.maxZirCAD CEREC/in Lab MT Multi, CE= CeramillZolidfx ML. cg = control glazed, cp = control polished, c = cervical, m = medium, i = incisal.

Among glazed tested groups, the lowest mean Ra values were recorded for ST (0.064 to 0.067) and the highest for CE (0.086 to 0.084), with very close values between layers. For IP, the values tend to increase from the cervical (0.073 ± 0.009) to incisal (0.083± 0.009) area.

Among the polished tested groups, the lowest Ra values were measured for ST (0.041 to 0.043) and the highest for CE (0.045 to 0.047), which were almost similar in all three areas. For IP, an upward trend was registered, the lowest value (0.043 ± 0.010) was in the cervical area and the highest (0.046 ± 0.009) in the incisal one.

Mean values and SD (standard deviation) for Ra (µm)-arithmetic average surface roughness and Rz (µm)-maximum surface roughness measurements after aging are summarized in Table 3.

Table 3. Mean values with SD for Ra, Rz parameters of the samples, after aging (n = 8).

		ST ag	ST ap	IP ag	IP ap	CE ag	CE ap
	c	0.080 ± 0.009	0.054 ± 0.010	0.081 ± 0.009	0.052 ± 0.009	0.088 ± 0.009	0.048 ± 0.008
Ra	m	0.078 ± 0.009	0.055 ± 0.010	0.084 ± 0.008	0.050 ± 0.009	0.087 ± 0.009	0.047 ± 0.009
	i	0.077 ± 0.009	0.052 ± 0.010	0.086 ± 0.010	0.049 ± 0.008	0.086 ± 0.009	0.048 ± 0.009
	c	0.535 ± 0.100	0.426 ± 0.101	0.543 ± 0.094	0.372 ± 0.090	0.576 ± 0.099	0.413 ± 0.095
Rz	m	0.520 ± 0.095	0.422 ± 0.099	0.548 ± 0.097	0.383 ± 0.091	0.579 ± 0.095	0.406 ± 0.091
	i	0.518 ± 0.092	0.410 ± 0.091	0.562 ± 0.089	0.390 ± 0.094	0.591 ± 0.084	0.416 ± 0.090

Ra-arithmetic average surface roughness, Rz-maximum surface roughness. ST= STML, IP= IPS e.maxZirCAD CEREC/in Lab MT Multi, CE= CeramillZolidfx ML. ag = aged glazed, ap = aged polished, c = cervical, m = medium, i = incisal.

Among glazed tested groups, it was observed that the surface roughness increased in all areas for the three materials, the lowest mean Ra values were recorded for ST (0.077 to 0.080) and the highest for CE (0.086 to 0.088), with very close values between layers. For IP, the Ra values increased significantly in the cervical area and slightly in incisal area; however, the surface roughness is lower in cervical area (0.081 ± 0.009) than in incisal area (0.086 ± 0.010) for this material.

Among polished tested groups the lowest Ra values were measured for ST (0.052 to 0.054) and the highest for CE (0.047 to 0.048), almost similar in all three areas. For IP, the values tend to decrease from the cervical (0.053 ± 0.009) to incisal (0.049 ± 0.009) area.

For the glazed side, the lowest Rz values was recorded for ST (0.512 to 0.435) and the highest (0.576 to 0.591) for CE and for IP an intermediate appearance with the lowest value (0.534 ± 0.094) in cervical area and the highest (0.562 ± 0.089) in incisal ones.

On the polished side, the lowest Rz values (0.406 to 0.416) were reported for CE and the highest (0.518 to 0.535) for ST.

Mean roughness measurements performed before and after aging, in three areas (cervical, medium, incisal) on the glazed and polished samples (n = 8) are presented in Figure 2.

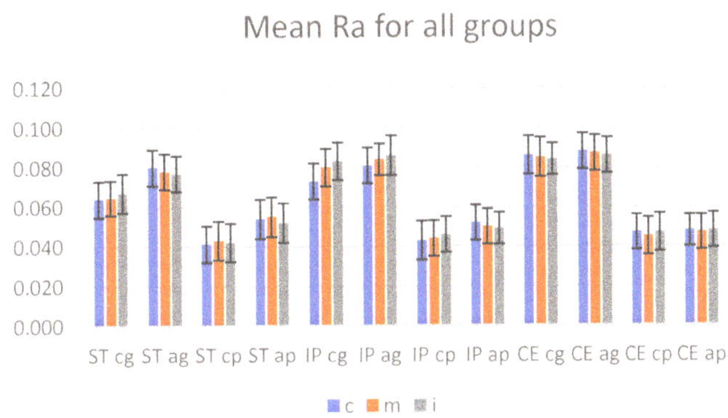

Figure 2. Mean Ra values (arithmetic average surface roughness) with SD (standard deviation), before and after aging for all groups. ST = STML, IP = IPS e.maxZirCAD CEREC/in Lab MT Multi, CE = CeramillZolidfx ML. cg = control glazed, cp = control polished, ag = aged glazed, ap = aged polished, c = cervical, m = medium, i = incisal.

Before aging, the two-way ANOVA test ($\alpha = 0.05$) reported insignificant differences ($p > 0.05$) when the statistical test was performed to compare the surface roughness between cervical-medium-incisal areas individual on glazed (for Ra: ST $p = 0.771$, IP $p = 0.812$, CE $p = 0.909$; for Rz: ST $p = 0.504$, IP $p = 0.587$, CE $p = 0.704$), respectively on polished samples (for Ra: ST $p = 0.814$, IP $p = 0.692$, CE $p = 742$; for Rz: ST $p = 0.784$, IP $p = 0.831$, CE $p = 0.863$); also after aging, no significant differences ($p > 0.05$, $\alpha = 0.05$) were observed when comparing the every three areas, on each glazed or polished surface, for a material.

Both before and after aging, the surface roughness values for the glazed samples are higher than for those polished. The unpaired *t*-Test test ($\alpha = 0.05$) reported a significant difference ($p < 0.05$) in terms of Ra-arithmetic average surface roughness and Rz-arithmetic average surface roughness betweenthe glazed and polished groups in the same areas, for all three materials; the *p* values are included in Table 4.

Table 4. The unpaired *t*-Test test ($\alpha = 0.05$)-*p* values regarding Ra, Rz for control, glazed-control polished, respectively aged glazed-aged polished groups in each area.

		ST cg-cp	IP cg-cp	CE cg-cp	ST ag-ap	IP ag-ap	CE ag-ap
Ra	c	0.001	<0.001	<0.001	0.002	<0.001	<0.001
	m	0.004	<0.001	<0.001	<0.001	<0.001	<0.001
	i	<0.001	<0.001	<0.001	<0.001	<0.001	<0.001
Rz	c	0.039	0.026	0.004	0.042	0.005	0.006
	m	0.044	0.009	0.006	0.039	0.005	0.003
	i	0.031	0.005	0.002	0.046	0.003	0.001

Ra-arithmetic average surface roughness, Rz-maximum surface roughness. ST = STML, IP = IPS e.maxZirCAD CEREC/in Lab MT Multi, CE = CeramillZolidfx ML. cg = control glazed, cp = control polished, ag = aged glazed, ap = aged polished. c = cervical, m = medium, i = incisal.

Although it has been observed that the surface roughness has increased on polished, respectively on glazed side after hydrothermal aging, the statistical paired *t*-Test ($\alpha = 0.05$) reported significant differences concerning Ra, Rz among the samples before—after autoclaving only for ST, the *p* values are included in Table 5.

Table 5. The paired *t*-Test test (α = 0.05)-*p* values regarding Ra, Rz for control glazed–agedglazed, respectively control polished–aged polished groups in each area.

		ST cg-ag	ST cp-ap	IP cg-ag	IP cp-ap	CE cg-ag	CE cp-ap
	c	0.006	0.049	0.134	0.058	0.715	0.868
Ra	m	0.046	0.022	0.521	0.387	0.662	0.628
	i	0.042	0.047	0.583	0.442	0.311	0.808
	c	0.371	0.225	0.775	0.621	0.583	0.866
Rz	m	0.412	0.425	0.621	0.539	0.715	0.754
	i	0.344	0.568	0.592	0.546	0.662	0.862

Ra-arithmetic average surface roughness, Rz-maximum surface roughness. ST = STML, IP = IPS e.maxZir CAD CEREC/in Lab MT Multi, CE = CeramillZolidfx ML. cg = control glazed, cp = control polished, ag = aged glazed, ap=aged polished. c = cervical, m = medium, i = incisal.

3.2. Atomic Force Microscopy (AFM)

Before aging, for each kind of material, fine and deeper interconnecting striations and scratches, small irregularities as scattered and sharp peaks were observed on both treated surfaces in all three areas. Among materials, for CE, the surface irregularities are more obvious both in the group of glazed and those of polished samples.

The values for parameters Sa (nm)-arithmetical mean height and Sq (nm)-root mean square height resulted for the glazed and polished samples are showed in Table 6.

Table 6. Sa, Sq values for control-glazed and control-polished samples, in each area (n = 8).

		ST cg	IP cg	CE cg	ST cp	IP cp	CE cp
	c	1.651	1.960	1.723	1.048	1.244	1.388
Sa	m	2.424	2.536	2.662	1.315	1.381	1.582
	i	1.524	1.598	1.924	1.529	1.529	1.343
	c	1.934	2.293	1.981	1.348	1.335	1.738
Sq	m	2.945	3.049	3.123	1.691	1.552	2.113
	i	1.842	1.996	2.144	1.945	2.045	1.545

Sa = arithmetical mean height, Sq = root mean square height. ST = STML, IP = IPS e.maxZirCAD CEREC/in Lab MT Multi, CE = CeramillZolidfx ML. cg = control glazed, cp = control polished, c = cervical, m = medium, i = incisal.

The AFM 3D images of the control group recorded on each area of glazed samples are shown in Figure 3.

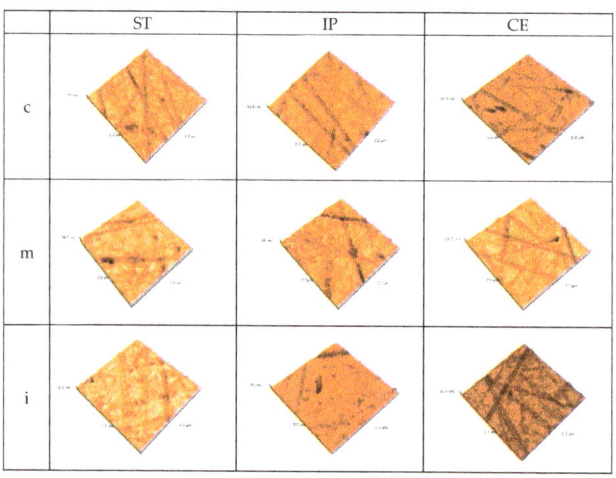

Figure 3. 3D AFM images of the glazed samples, before aging.ST = STML, IP = IPS e.maxZirCAD CEREC/in Lab MT Multi, CE = CeramillZolidfx ML. c = cervical, m = medium, i = incisal.

The AFM 3D images on each area of the polished samples of the control group are shown in Figure 4.

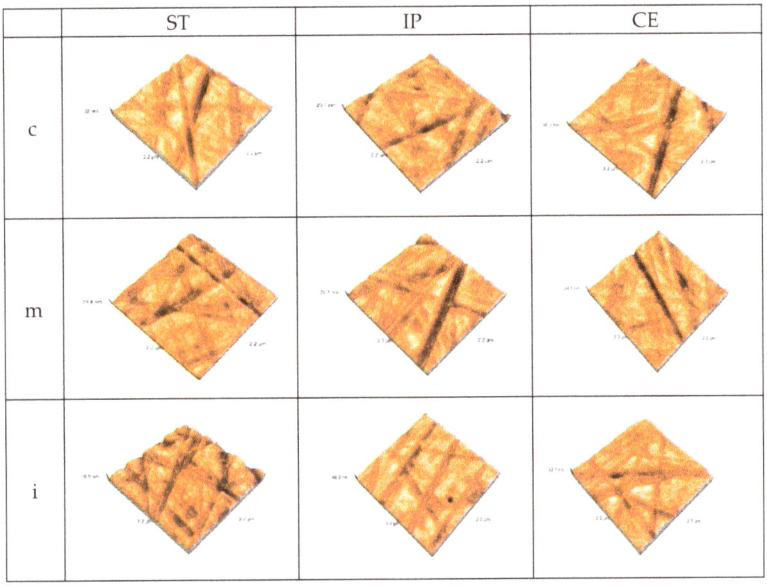

Figure 4. 3D AFM image of the polished samples of the control group. ST= STML, IP= IPS e.maxZirCAD CEREC/in Lab MT Multi, CE= CeramillZolidfx ML, c = cervical, m = medium, i = incisal.

After aging, the values for Sa (nm)-arithmetical mean height and Sq (nm)-root mean square height resulted for the glazed samples and polished samples are shown in Table 7.

Table 7. Sa, Sq values for aged-glazed and aged-polished samples, in each area (n = 8).

		ST ag	IP ag	CE ag	ST ap	IP ap	CE ap
	c	23.334	28.134	31.817	20.651	27.714	20.342
Sa	m	21.419	29.050	27.315	31.783	27.674	22.991
	i	22.458	24.262	24.464	28.978	25.502	20.515
	c	26.744	32.526	36.371	23.669	31.778	23.529
Sq	m	24.414	33.221	31.242	36.254	31.952	25.883
	i	25.702	27.911	27.873	32.988	29.106	23.728

Sa = arithmetical mean height, Sq = root mean square height. ST = STML, IP= IPS e.maxZirCAD CEREC/in Lab MT Multi, CE = CeramillZolidfx ML. ag = aged-glazed, ap = aged-polished, c = cervical, m = medium, i = incisal.

After aging, the AFM 3D images recorded the topography on each area of the glazed samples—Figure 5.

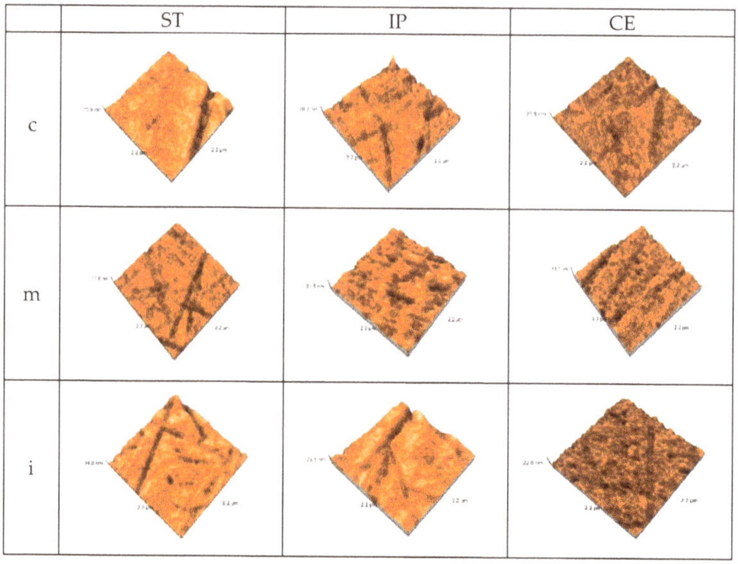

Figure 5. 3D AFM images of the glazed samples, after aging.ST = STML, IP = IPS e.maxZirCAD CEREC/in Lab MT Multi, CE = CeramillZolidfx ML c = cervical, m = medium, i = incisal.

After autoclaving, the AFM 3D images on each area for polished samples are represented in Figure 6.

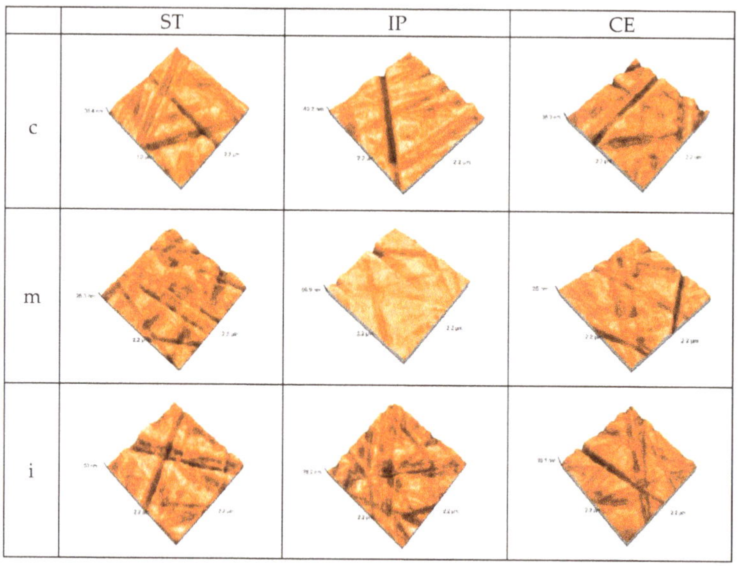

Figure 6. 3D AFM images of the polished samples, after aging.ST= STML, IP= IPS e.maxZirCAD CEREC/in Lab MT Multi, CE = CeramillZolidfx ML, c = cervical, m = medium, i = incisal.

After aging, there is a significant increase in Sa (nm)-arithmetical mean height and Sq (nm)-root mean square height values, for all materials, both on the glazed and polished surfaces.

On the glazed samples Sa (nm) values have an ascending aspect from ST to CE. ST in the medium area recorded the lowest value 21.41 ± 0.02 than 2.42 ± 0.01-before; the highest value resulted for CE in cervical area 31.81 ± 0.01, compared to 1.72 ± 0.01-before. For IP in medium area was reported an intermediate value 29.05 ± 0.02 compared to 2.54 ± 0.01-before.

On the polished samples, the lowest value was reported for CE in cervical area 20.34 ± 0.01, compared to 1.39 ± 0.01, before and the highest for ST in incisal and medium area 31.78 ± 0.02 than 1.32 ± 0.02, before. For IP, the values have an intermediate aspect, but they are almost similar to those on glazed side.

On glazed side of specimens, the lowest Sq value is in medium area for ST-24.41 ± 0.02 compared to 2.94 ± 0.01, before and the highest for CE in incisal area 36.37 ± 0.01, compared to 1.98 ± 0.01, before. For IP in medium area Sq value 33.22 ± 0.02 was recorded, compared to 3.05 ± 0.01, before.

On the polished side of samples, the lowest Sq values werereported for CE in cervical area 23.52 ± 0.01, compared to 1.74 ± 0.01, before and the highest for ST in medium area 36.25 ± 0.02 than 1.69 ± 0.02, before.

Before aging, the two-way ANOVA test ($\alpha = 0.05$) reported no significant difference ($p > 0.05$) regarding to Sa = root mean square height, Sq = root mean square height, between the three areas (cervical-medium-incisal) of glazed or polished samples for a material.

The unpaired *t*-Test test ($\alpha = 0.05$) reported significant differences ($p < 0.05$) in terms of Sa-arithmetical mean height in cervical and medium areas and no significant differences in incisal area between glazed and polished surfaces of each material (incisal: ST glazed-polished $p = 0.08$; IP glazed-polished $p = 0.241$; CE glazed-polished $p = 0.452$).

After aging, the two-way ANOVA test ($\alpha = 0.05$) reported no significant difference ($p > 0.05$) regarding to Sa = root mean square height, Sq = root mean square height, between the three areas(cervical, medium, incisal) for each glazed or polished samples, of a material and the unpaired *t*-Test test ($\alpha = 0.05$) reported significant differences ($p < 0.05$) in terms of Sa-arithmetical mean height, between the same areas of the glazed and polished surfaces of each material, for ST in cervical and medium area.

The before–after aging statistical paired *T*-Test ($\alpha = 0.05$) found significant differences ($p \leq 0.05$) regarding Sa, Sq values among the same areas of glazed, respectively polished samples.

3.3. Scanning Electron Microscopy (SEM)

The surface topographic analysis revealed a regular and homogeneous structure for the glazed samples, before aging, except some places with a wavy surface due to the application of glaze. In the medium area on ST followed by IP samples, the surface displayed interconnecting striations and was rougher than the other areas; by IPS and CE in incisal area, fine striations are visible under glaze layer.

On the polished samples, before aging, the striations, small irregularities, and defects are more obvious than on the glazed surfaces; for ST and IP in medium area, flaws and porosities were observed on the surface.

SEM images of the control samples are illustrated in Figures 7 and 8.

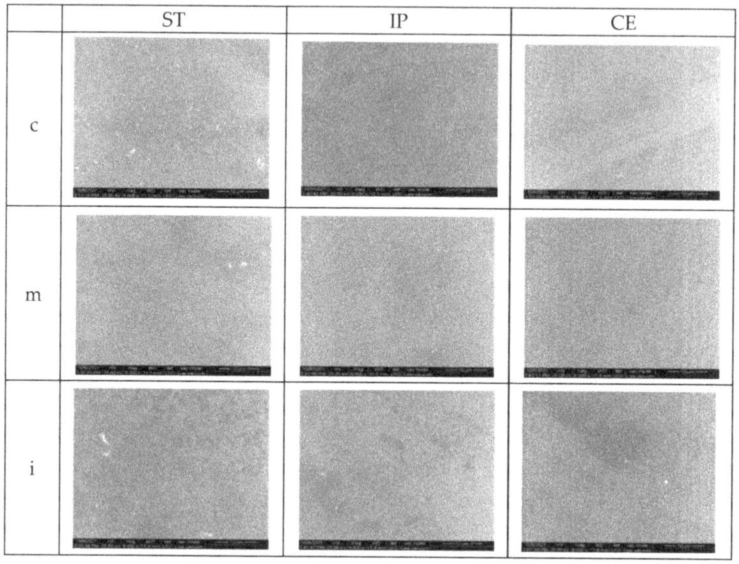

Figure 7. SEM images for glazed samples before aging. ST= STML, IP= IPS e.maxZirCAD CEREC/in Lab MT Multi, CE= CeramillZolidfx ML, c = cervical, m = medium, i = incisal.

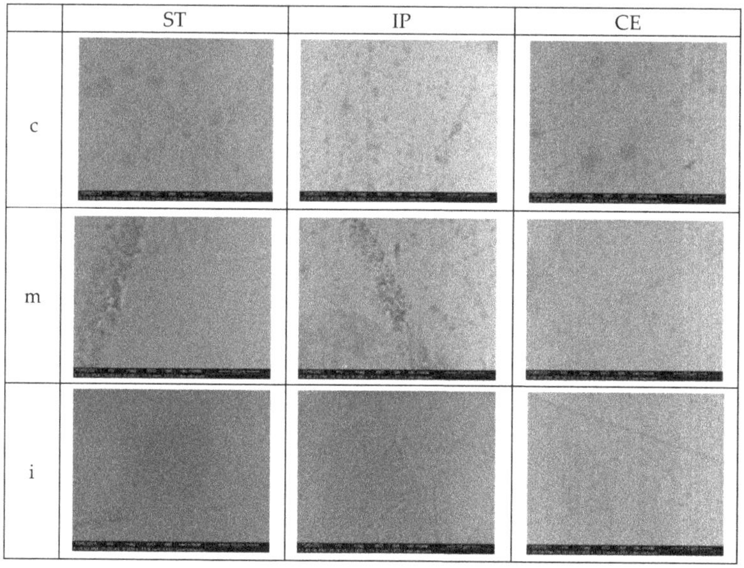

Figure 8. SEM images for polished samples before aging.ST= STML, IP= IPS e.maxZirCAD CEREC/in Lab MT Multi, CE= CeramillZolidfx ML, c = cervical, m = medium, i = incisal.

After aging for glazed specimens, the appearance of the surfaces is quite similar; however, on ST samples in cervical and medium area, striations and micro-cracks may be observed. After autoclaving, for CE, the appearance of the surface is quite similar to that before aging. On polished samples, there was an obvious damage in all areas compared to the situation before aging.

After aging, the SEM images of samples are illustrated in Figures 9 and 10.

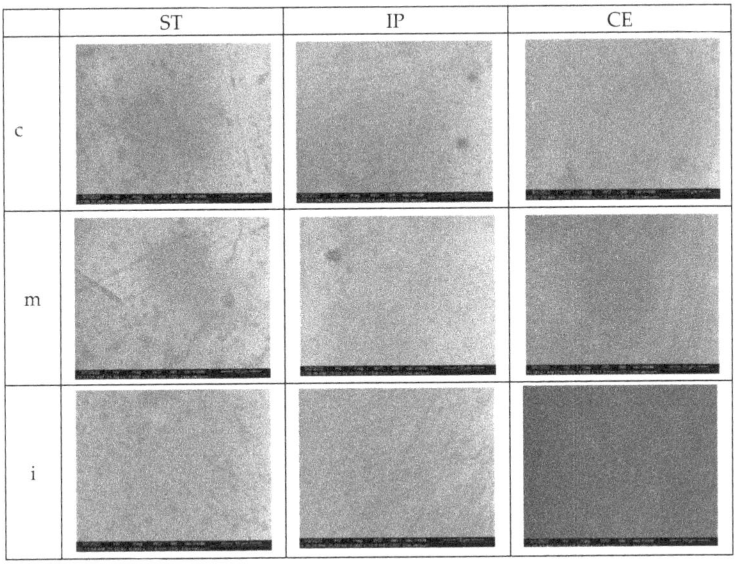

Figure 9. SEM images for glazed samples after aging.ST = STML, IP = IPS e.maxZirCAD CEREC/in Lab MT Multi, CE = CeramillZolidfx ML, c = cervical, m = medium, i = incisal.

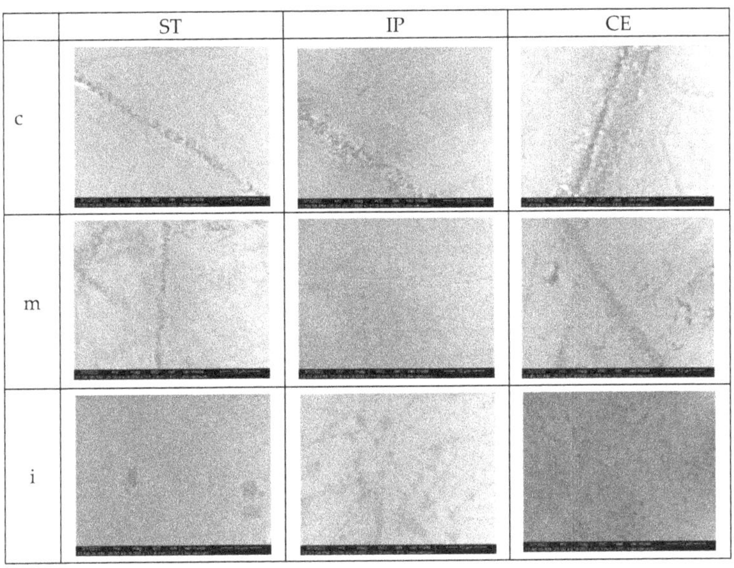

Figure 10. SEM images for polished samples after aging.ST = STML, IP = IPS e.maxZirCAD CEREC/in Lab MT Multi, CE = CeramillZolidfx ML, c = cervical, m = medium, i = incisal.

3.4. Microhardness Testing

Before and after aging, the mean Vickers hardness values with SD (standard deviation) of the materials, for both glazed and polished specimens in three areas are presented in Table 8.

Table 8. Mean Vickers hardness values with SD, for all samples (n = 8).

		ST Glazed	ST Polished	IP Glazed	IP Polished	CE Glazed	CE Polished
control groups	c	770.33 ±9.60	1763 ± 17.02	700 ± 5.72	1732 ± 15.70	632.67± 8.50	1637 ± 12.50
	m	763.67 ± 8.62	1763 ± 17.02	597 ± 3.80	1638.67 ± 14.02	617 ± 7.90	1552 ± 8.41
	i	694.67 ± 9.14	1732 ± 15.89	626 ± 4.20	1700.67 ± 15.03	610 ± 7.28	1597 ± 9.54
aged groups	c	672.67 ± 8.32	1711 ± 15.41	621 ± 4.10	1729.67 ± 15.05	708.33 ± 9.42	1532.33 ± 9.50
	m	601 ± 5.12	1809.67 ± 17.91	603.33 ± 4.08	1764.33 ± 16.15	722± 8.11	1607.67 ± 8.97
	i	640.33 ± 5.81	1797 ± 17.56	694.33 ± 4.21	1762.67 ± 16.08	722± 8.11	1667.67± 5.70

ST = STML, IP = IPS e.maxZirCAD CEREC/in Lab MT Multi, CE = CeramillZolidfx ML; c = cervical, m = medium, i = incisal.

On the glazed samples, the highest microhardness values were reported for ST (770.33 ± 9.60), followed by IP (700 ± 5.72) and CE (632.67 ± 8.50) for all, in the cervical area. On the polished samples, the highest surface microhardness was recorded for ST in cervical area and medium area (1763 ± 17.02), followed by IP in cervical area (1732 ± 15.70) and CE in cervical area (1637 ± 12.50).

Among materials, for CE, on the glazed and polished surface, the microhardness values were lower.

After aging, the microhardness for polished samples issignificantly higher compared to the glazed ones.

On the glazed side, for ST, the values decreased, the highest was recorded (672.67 ± 8.32) in cervical area; for IP the values are almost the same as before aging (ex. 694.33 ± 4.21) in incisal area. For CE, an increase in all three areaswas reported; the highest values (722 ± 8.11) were recorded in medium and incisal area.

On the polished samples, a minor increase in values can be observed; the highest (1667 ± 5.70) was reported for CE in incisal area.

The microhardness values recorded before and after aging on glazed and polished samples, in cervical, medium, incisal areas, for all three materials (n = 8), are summarized in Figure 11.

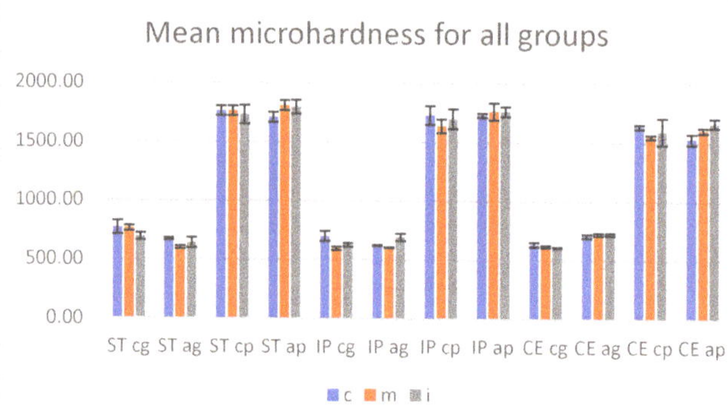

Figure 11. Mean microhardness values and SD before and after aging. ST = STML, IP = IPS e.maxZirCAD CEREC/in Lab MT Multi, CE = CeramillZolidfx ML; cg = control glazed, cp = control polished, ag = aged glazed, ap = aged polished;c = cervical, m = medium, i = incisal.

Before aging, the microhardness values for polished specimens were significantly higher compared to the glazed ones.

There were insignificant differences ($p > 0.05$) when the statistical test two-way ANOVA was performed to compare the surface microhardness between cervical-medium-incisal areas individual on glazed and on polished sides of each zirconia.

The unpaired *t*-Test test ($\alpha = 0.05$) assessed a significant impact ($p < 0.005$) to the analysis in the same areas between control-glazed and control-polished groups, for all three materials.

For both glazed and polished surfaces, Pearson's correlation reported a negative correlation between surface roughness Ra and microhardness, $r = -0.9$, which means that the hardness decreases when the roughness is increased.

After autoclaving, no significant differences ($p > 0.05$) were observed by statistical two-way ANOVA test between every two or three areas, on each glazed or polished surface for a material.

The unpaired *t*-Test test ($\alpha = 0.05$) revealed significant differences ($p < 0.005$) to the analysis in the same areas between aged-glazed and aged-polished groups, for all three materials.

The statistical paired *t*-Test ($\alpha = 0.05$) reported no significant differences concerning microhardness values (Vickers Hardness number-VN) among the same areas recorded before and after aging, on the glazed and polished surfaces, except CE glazed samples; the *p* values are included in Table 9.

Table 9. The paired *t*-Test test ($\alpha = 0.05$)-*p* values for microhardnessamong control glazed-aged glazed, respectively control polished-aged polished groups in each area.

VN	ST cg-ag	STcp-ap	IP cg-ag	IP cp-ap	CE cg-ag	CE cp-ap
c	0.621	0.033	0.887	0.972	0.012	0.091
m	0.116	0.248	0.584	0.623	0.016	0.183
i	0.413	0.475	0.671	0.523	0.001	0.359

VN = Vickers Hardness number (microhardness value). ST = STML, IP = IPS e.maxZirCAD CEREC/in Lab MT Multi, CE = CeramillZolidfx ML. cg = control glazed, cp=control polished, ag = aged glazed, ap=aged polished. c = cervical, m = medium, i = incisal.

4. Discussion

The study investigated the surface characteristics, microstructure and microhardness on glazed and polished samples of three multilayered dental zirconia with different percentages of Y_2O_3 (4 mol%; 4 mol%+ 5 mol%; 5 mol%), first before and then after aging, achieved through autoclaving, as LTD (low temperature degradation).

Before aging, among materials ST = STML, IP = IPS e.maxZir CAD CEREC/in Lab MT Multi, CE = CeramillZolidfx ML for the glazed samples, the lowest Ra-arithmetic average surface roughness values were found for ST (4 mol% Y_2O_3), followed by IP (4 mol% + 5 mol% Y_2O_3) and the highest surface roughness was recorded for CE (5 mol% Y_2O_3)-the superhigh-translucency zirconia, with the highest Yttrium content and the lowest flexural strength (700 MPa) of the three analyzed materials.

Among areas of the glazed samples, for ST, the highest Ra value (0.067 ± 0.010) was recorded in the incisal area; for IP, it was found that surface roughness increases from cervical (0.073 ± 0.009) to incisal (0.083 ± 0.009); and for CE close values were reported, the highest cervical (0.086 ± 0.010). Between the glazed (cervical-medium-incisal) areas of a material no statistically significant differences were found ($p < 0.05$). On the polished samples, the same ratio was observed as for glazed specimens, the lowest Ra values were found for ST (ex. 0.043 ± 0.010—medium), followed by IP (ex. 0.046 ± 0.009-incisal) and the highest for CE (ex.0.047 ± 0.008-cervical). Among (cervical-medium-incisal) areas on glazed or polished surfaces of a material no statistically significant differences (two-way ANOVA test, $\alpha = 0.05$) were found ($p < 0.05$).

The surface roughness values measured for the glazed samples are higher than for those polished. The unpaired *t*-Test ($\alpha = 0.05$) reported a significant difference ($p < 0.05$) in terms of Ra and Rz between the same areas of glazed and polished sample, for all three materials.

When comparing the surface roughness obtained between the areas of a material, it is observed that for ST, respectively for CE, the values are close to each other, both on glazed and polished samples; which means as some manufactures and some studies show, the layers presented almost the same content of Yttrium, cubic phase, respectively a similar microstructure; the color difference being given only by the pigment composition [24]. Instead, for IP, the surface roughness values reported an ascending orderfrom cervical to incisal, corresponding to the various distribution of Yttrium content and cubic phase, respectively, and a different microstructure among the layers of this material.

Regarding glazing process as a surface treatment, the microstructure, the smoothness of the surface on which the glaze layer is deposited, and the interface between them greatly influence the external appearance of the surface. For ST, which contains smaller grains than CE, the surface roughness was lower. It is also possible to record different Ra, Rz values when using the manufacturer's own glaze pasta.

Referring to the polishing process, for this study small-grit grinding tools were used so as not to cause, as much as possible, phase transformation and surface degradation which results in an increase in surface roughness. However, a zirconia such as CE, that has the lowest mechanical properties among the studied materials and contains larger grains that can be easily dislocated during processing resulting unevenness and flaws, presented the highest surface roughness.

In addition, for a better surface topography evaluation, the samples were examined using an atomic force microscope (AFM) and a scanning electron microscope (SEM).

AFM provided a high resolution three-dimensional visualization of glazed and polishing samples, the values for parameters Sa (nm)-arithmetical mean height and Sq (nm)-root mean square height resulted. The nanoroughness values found on the glazed surfaces were higher than on the polished ones. On the glazed samples, the highest Sa value was recorded for CE (2.66 $\pm$ 0.02), followed by IP (2.54 $\pm$ 0.01) and the least affected was ST (2.42 $\pm$ 0.01), all in the medium area. On the polished surfaces, the highest Sa (nm) value resulted for CE (1.58 $\pm$ 0.01) in medium area, IP (1.55 $\pm$ 0.01), and almost the same for ST (1.53 $\pm$ 0.02) in incisal area.

The SEM examination for glazed samples revealed a fairly smooth and uniform surface with striated patterns resulting from the deposition (brushing) of the glaze layers, creating a wavy appearance, which might influence the surface roughness. By CE and IP-incisal areas, the striations resulting from processing are visible under the glaze layers. The images which present the polished surfaces showed interconnecting scratches, shallow striation and micro-cracks created by the rotative instrument on a smooth surface, except for ST and IP medium area, where grater damage was observed.

Surface microhardness test measures the resistance of material to successive indentation. Mechanical properties are influenced by the crystalline phase, yttrium content, more precisely by the microstructure of the material [52]. In this study, the unpaired *t*-Test test ($\alpha = 0.05$) reported a significant difference ($p < 0.005$) regarding the microhardness (VN) between polished (1552 to 1763 MPa) and glazed surfaces (597 to 763 MPa).

The glaze layer was locally negatively damaged; thus confirming that the surface treatment can affect the surface hardness and the resistance of the material. The sharpest and most chipped edge of the indent was observed on the glazed surface of CE (5 mol% Y_2O_3), followed by ST (4 mol% Y_2O_3) and the least obvious crack patterns and indentation imprints by IP (4 mol% + 5 mol% Y_2O_3).

Considering that the same glaze was used for all three materials, it is possible that the microstructure, surface appearance, the interface had an influence on the sealing, adhesion capacity, and resistance of the glaze layer. Moreover, it may be relevant as the glaze (IPS Ivocolor Glaze Paste) and IP (IPS e.maxZirCAD CEREC/in Lab MT Multi) materials were produced by the same manufacturer (IvoclarVivadent, Schaan, Liechtenstein). Among the three areas of glazed, respectively of polished surfaces, no significant differences (two-way ANOVA test, $p > 0.05$) were reported, confirming the first hypothesis, also in terms of hardness.

For both glazed and polished specimens, a negative Pearson's correlation (r = -0.9) between surface roughness and microroughness was reported; meaning that increasing the roughness implies decreasing the hardness and the surface treatment significantly influenced the surface roughness, respectively the microhardness [53].

After aging, both glazed and polished samples show an increase in roughness and the values recorded on the glazed samples were higher than for those polished ones.

Among glazed specimens, compared with the situation before aging, statistical paired *t*-Test (α =0.05) revealed a significant increase in values for ST ($p < 0.05$) and an insignificant increase for CE ($p > 0.05$); however, the lowest mean Ra values were recorded for ST (0.077 to 0.080) and the highest for CE (0.086 to 0.088). For IP, the surface roughness increased in cervical area and slightly in incisal area, but the values were lower in cervical area (0.081 $\pm$ 0.009) than in incisal area (0.086 $\pm$ 0.010).

As a result, the conclusion could be that the material with 4 mol% Yttrium content (ST, IP–cervical) undergoes a more intense aging process accompanied by phase transformation, volume increase of particles, grain detachment, and surface roughness. For CE (5 mol% Yttrium), the surface roughness before aging was the highest and after aging showed a minor increase, meaning that the superhigh-translucency zirconia is not very affected by aging; however, it remains the material with the roughest surface.

Other studies demonstrated as well that the increased roughness does not accelerate the aging process for (super) high-translucent zirconia and the toughness was preserved after grinding and LTD [16,54].

Certainly, the glaze layer has a protective role, but may be directly affected by the aging process or as a result of alteration in the underlying material, with influence on surface texture and hardness.

Among polished samples, compared with the situation before aging, the paired *t*-Test (α = 0.05) reported that the surface roughness values for ST ($p < 0.005$) increased more than those for CE ($p > 0.005$) and among materials, the highest Ra values were measured for ST (0.052 to 0.054) and the lowest for CE (0.047 to 0.048); for IP, the values tend to decrease from the cervical (0.053 $\pm$ 0.009) to incisal (0.049 $\pm$ 0.009) area. This means that the material with 4 mol% Yttrium content is more susceptible to the aging process, thus hydrothermal and mechanical impact cause phase transformation, resulting in surface degradation and an increase in microhardness; this situation has been presented in other studies [55,56].

On the glazed surfaces, the nanoroughness values Sa = arithmetical mean height, Sq = root mean square height, were higher than on the polished ones;the lowest values were found for ST and the highest for CE. On the polished specimens, the highest nanoroughness values were recorded for ST and the lowest for CE. Regarding Sa and Sq, the paired *t*-Test (α = 0.05) reported a significant increase in glazed and polished specimens, after aging.

After aging, the surface topographic analysis (SEM) revealed for ST glazed specimens, more visible striations, and micro-cracks (cervical, medium- area), compared to the appearance before aging. For CE, a minor increase in surface roughness was recorded and the surface aspect is quite similar to that before aging. On polished samples compared to the initial situation, in all areas, an evident degradation with deeper striations and scratches, sharp peaks, irregularities and defectswas observed.

It was revealed that both before and after autoclaving, glazing and polishing as surface treatment, influenced and affected the microroughness and topography of zirconia with different Yttrium content. The second null hypothesis is accepted.

Although an optimal threshold has not been set, it was reported that the roughness value (Ra-arithmetic average surface roughness) above 0.2 μm, could induce negative effects such as retention of bacteria, periodontal inflammation, dental caries risk [57], and a value above 0.3 μm can be perceived by the tongue, negatively affecting patient comfort [57,58]. The Ra values results of this study, for all three zirconia types, in three areas on glazed, respectively polished surfaces, were lower than 0.2 μm, therefore they may be considered clinically acceptable.

Regarding microhardness, after autoclaving, the unpaired *t*-Test test (α = 0.05) revealed that the values for polished samples are maintained significantly higher ($p < 0.005$) compared to the glazed one, confirming the second hypothesis, also in terms of hardness.

On the glazed specimens, for ST was observed a decrease in microhardness, while those for IP remained at a level similar to that before aging and CE, denotingan increase in values. On the polished samples, for ST and IP a minor increase of values was recorded and for CE the values remain approximately the same as those before aging.

Before aging, ST (4 mol% Yttrium) recorded the lowest surface roughness and the highest microhardness, demonstratingthe best mechanical properties among the studied materials, on the glazed and polished samples. After aging, this material experienced the largest increase in roughness for both groups and microhardness decreased on glazed and increased on polished side. It seems that for the glazed samples, the aging process induced t-to-m phase transformation with an increase in surface roughness, adversely affecting the interface features and the glaze layer and for the polished specimens, the phase transformation led to a toughening reaction.

For CE (5 mol% yttrium), before autoclaving, the microhardness values were close to those for IP, but lower than those measured for ST on glazed surfaces and the lowest among materials on polished samples. After aging, the microhardness increased for the glazed group and between materials was the highest; for the polished samples the values are maintained (with a slight increase) as before aging and among materials were the lowest. It seems that the material was not very affected by aging process, however the hardness increased and the glaze layer had a protective role.

For IP (4 mol% + 5 mol%), regarding mechanical properties an intermediate aspect among the tested materials, both on glazed and polished samples, before or after autoclavingwas observed. The records revealed similarities between properties and values measured in cervical area with those of ST and the incisal area with those of CE.

The aging behavior is correlated with the microstructure of the material, and with the surface treatment. As a result of these findings, the third null hypotheses—the roughness value would increase and microhardness decrease after the aging procedure—is not accepted.

The limitation of this study was the short-term aging and the fact that certain factors that interact with restoration in oral environment could not be fully reproduced (chewing force, saliva components, temperature, and pH fluctuations).

5. Conclusions

1. For tested zirconia materials, no significant differences among layers were registered regarding surface characteristics.
2. Surface treatment (glazing or polishing) has a significant impact on surface roughness and microhardness; the glazed samples were found with higher surface roughness and lower microhardness compared to the polished ones. This correlation is preserved even after LTD.
3. After aging microroughness increases, but significant only for the super translucent material. There is a significant increase in nanoroughness values, for all materials, both on the glazed and polished surfaces. Related to microhardness, aging does not have a significant influence.
4. The super translucent 4Y-TZP (fourth-generation) material was more affected by aging compared to the super-high translucent 5Y-TZP (third-generation) material. The material with 4 mol% + 5 mol% revealed similarities for each layer corresponding to the microstructure.

Author Contributions: Conceptualization, F.R.T. and L.P.; methodology, F.R.T. and L.P.; software, F.R.T. and L.C.M.; validation, F.R.T. and L.P.; investigation, F.R.T., M.I.B., I.-D.U. and R.D.V.; writing—F.R.T.; writing—review and editing, L.P.; supervision, L.P. All authors have read and agreed to the published version of the manuscript.

Funding: This research received no external funding.

Institutional Review Board Statement: Not applicable.

Informed Consent Statement: Not applicable.

Data Availability Statement: Not applicable.

Conflicts of Interest: The authors declare no conflict of interest.

References

1. Denry, I.; Kelly, J.R. State of the art of zirconia for dental applications. *Dent. Mater.* **2008**, *24*, 299–307. [CrossRef] [PubMed]
2. Yamada, Y.; Matsumoto, A.; Ito, Y. Zirconia Sintered Body, Zirconia Composition and Zirconia Calcined Body Production Method. U.S. Patent 9212065, 15 December 2015.
3. Salmang, H.; Scholze, H. *Keramik Volständing Neuarbaitete und Erweiterte Auflage;* Springer: Heidelberg, Germany, 2007; Volume 7, pp. 817–838.
4. Holleman-Wiberg, E. *Lehrbuch der Anorganischen Chemie*, 101st ed.; Walterde Gruyter & Co.: Berlin, Germany, 1995; pp. 918–985.
5. Zhang, Y.; Inokoshi, M.; Batuk, M.; Hadermann, J.; Naert, I.; Van Meerbeek, B.; Vleugels, J. Strength, toughness and aging stability of hightly translucent Y-TZP ceramics for dental restorations. *Dent. Mater.* **2016**, *32*, 327–337. [CrossRef] [PubMed]
6. Zhang, Y.; Lawn, B.R. Novel zirconia materials in dentistry. *J. Dent. Res.* **2018**, *97*, 140–147. [CrossRef] [PubMed]
7. Lawson, N. Everything You Need to Know about Dental Zirconia. Available online: https://www.DentalAcademyofCE.com (accessed on 16 February 2022).
8. Scientific Documentation. Microsoft Word–WissDok IPS e.maxZirCADenglisch_aktualisiert 30112017.docx. Available online: https://www.ivoclar.com/ (accessed on 3 March 2022).
9. Zhang, F.; Reveron, H.; Spies, B.C.; Van Meerbeek, B.; Chevalier, J. Trade-off between fracture resistance and translucency of zirconia and lithium-desilicated glass ceramics for monolithic restorations. *Acta Biomater.* **2019**, *91*, 24–34. [CrossRef] [PubMed]
10. Bressan, E.; Ferroni, L.; Gardin, C.; Bellin, G.; Sbricoli, L.; Sivolella, S.; Brunello, G.; Schwartz-Arad, D.; Mijiritsky, E.; Penarrocha, M.; et al. Metal Nanoparticles Released from Dental Implant Surfaces: Potential Contribution to Chronic Inflammation and Peri-implant Bone Loss. *Materials* **2019**, *12*, 2036. [CrossRef]
11. Tatullo, M.; Genovese, F.; Aiello, E.; Amantea, M.; Makeeva, I.; Zavan, B.; Rengo, S.; Fortunato, L. Phosphorene is the Neu Graphene in Biomedical Application. *Materials* **2019**, *12*, 2301. [CrossRef]
12. Tatullo, M.; Zavan, B.; Genovese, F.; Codispoti, B.; Makeeva, I.; Fortunato, L.; Spagnuolo, G. Borophene Is a Promising 2D Allotropic Material for Biomedical Devices. *Appl. Sci.* **2019**, *9*, 3446. [CrossRef]
13. Garvie, R.C.; Hannink, R.H.; Pascoe, R.T. Ceramic steel? *Nature* **1975**, *258*, 703–704. [CrossRef]
14. Chevalier, J.; Cales, B.; Drouin, J.M. Low-temperature aging of Y-TZP ceramics. *J. Am. Ceram. Soc.* **1999**, *82*, 2150–2154. [CrossRef]
15. Kim, J.W.; Covel, N.S.; Guess, P.S.; Rekow, E.D.; Zhang, Y. Concerns of hydrothermal degradation in CAD/CAM zirconia. *J. Dent. Res.* **2010**, *89*, 91–95. [CrossRef]
16. Pereira, G.K.R.; Venturini, A.B.; Silvestri, T.; Dapieve, K.S.; Montagner, A.F.; Soares, F.Z.M.; Valandro, L.F. Low-temperature degradation of Y-TZP ceramics: A sytematic review and meta-analysis. *Mech. Behav. Biomed. Mater.* **2015**, *55*, 151–163. [CrossRef] [PubMed]
17. Kelly, J.; Denry, I. Stabilized zirconia as a structural ceramic: An overview. *Dent. Mater.* **2008**, *24*, 299–307. [CrossRef] [PubMed]
18. Deville, S.; Gremillard, L.; Chevalier, J.; Fantozzi, G. A critical comparison of methods for the determination of aging sensitivity in biomedical grade yttria-stabilized zirconia. *J. Biomed. Mater. Res. Part B Appl. Biomater.* **2005**, *72*, 239–245. [CrossRef] [PubMed]
19. Basililio, M.A.; Cardoso, K.V.; Antonio, S.G.; Rizialla, A.S.; Santos, J.G.C.; Arioli, F.J.N. Effects of artificial aging condition on yttria-stabilized zirconia implant abutsments. *J. Prosthet. Dent.* **2016**, *116*, 277–285. [CrossRef] [PubMed]
20. Harada, A.; Shishido, S.; Barkarmo, S.; Inagaki, R.; Kanno, T.; Örtengren, U.; Egusa, H.; Nakamura, K. Mechanical and microstructural properties of ultra-translucent zirconia ceramic stabilized with 5 mol% yttria. *J. Mech. Behav. Biomed. Mater.* **2020**, *111*, 103974. [CrossRef]
21. McLean, J.W. *The Science of Dental Ceramics*, 1st ed.; Quintessence Publishing Co.: Berlin, Germany, 1979; pp. 96–110.
22. Anselmi-Tamburini, U.; Woolman, I.N.; Munzir, Z.A. Transparent nanometric cubic and tetragonal zirconia obtaining by high-pressure pulsed electric current sintering. *Adv. Funct. Mater.* **2007**, *17*, 3267–3273. [CrossRef]
23. Porojan, L.; Vasiliu, R.D.; Birdeanu, M.I.; Porojan, S.D. Surface Characterisation of Dental Resin Composites Related to Conditioning and Finishing. *Polymers* **2021**, *13*, 4236. [CrossRef]
24. Kolakarnprasert, N.; Kaiser, M.R.; Kim, D.K.; Zhang, Y. New multi-layered zirconias: Composition, microstructure and translucency. *Dent. Mater.* **2019**, *35*, 797–806. [CrossRef]
25. Pereira, G.K.R.; Guilardi, L.F.; Dapieve, K.S.; Kleverlaan, C.J.; Rippe, M.P.; Valandro, L.F. Mechanical reliability, fatigue strength and survival analysis of new polycrystalline translucent zirconia ceramics for monolithic restorations. *J. Mech. Behav. Biomed. Mater.* **2018**, *85*, 57–65. [CrossRef]
26. Ueda, K.; Güth, J.F.; Erdelt, K.; Stimmelmayr, M.; Kappert, H.; Beuer, F. Light transmittance by a multi-colored zirconia material. *Dent. Mater. J.* **2015**, *34*, 310–314. [CrossRef]
27. Sedda, M.; Vichi, A.; Carrabba, M.; Capperucci, A.; Louca, C.; Ferrari, M. Influence of coloring procedure on flexural resistence of zirconia blocks. *J. Prosthet. Dent.* **2015**, *114*, 98–102. [CrossRef] [PubMed]
28. Aboushelib, M.N.; de Jager, N.; Kleverlaan, C.J.; Feilzer, A.J. The influence of pigments on the slow crack grow in dental zirconia. *Dent. Mater.* **2012**, *28*, 410–415. [CrossRef] [PubMed]

29. Kaiser, M.R.; Kokarnprasert, N.; Rodrigues, C.; Chai, H.; Zhang, Y. Probing the interfacial strength of novel multi-layer zirconias. *Dent. Mater.* **2020**, *36*, 60–67. [CrossRef] [PubMed]
30. Kim, H.K.; Kim, S.H.; Lee, J.B.; Han, J.S.; Yeo, I.S. Effect of polishing and glazing on the color and spectral distribution of monolithic zirconia. *J. Adv. Prosthodont.* **2013**, *5*, 296–304. [CrossRef] [PubMed]
31. Sarac, D.; Sarac, Y.S.; Yuzbasioglu, E. The effect of porcelain polishing systems on the color and surface texture of feldspathic porcelain. *J. Prosthet. Dent.* **2006**, *96*, 122–128. [CrossRef] [PubMed]
32. Deville, S.; Chevalier, J.; Gremillard, I. Influence of surface finish and residual stress on the aging sensitivity of biomedical grade zirconia. *Biomaterials* **2006**, *27*, 2186–2192. [CrossRef] [PubMed]
33. Alghazzawi, T.F.; Lemons, J.; Liu, P.R.; Essig, M.E.; Bartoluccci, A.A.; Janowski, G.M. Influence of low-temperature environmental exposure on the mechanical properties and structural stability of dental zirconia. *J. Prosthodont.* **2012**, *21*, 362–368.
34. Kohorst, P.; Borchers, L.; Strempel, J.; Stiesch, M.; Hassel, T.; Bach, F.W.; Hubsch, C. Low-temperature degradation of different zirconia ceramics for dental application. *Acta Biomater.* **2012**, *8*, 1213–1220. [CrossRef]
35. Canneto, J.; Cattani-Lorente, M.; Durual, S.; Wiskott, A.H.W.; Scherrer, S.S. Grinding damage assessment on four high-strength ceramics. *Dent. Mater.* **2016**, *32*, 171–182. [CrossRef]
36. Kosmac, T.; Oblak, C.; Jevnikar, P.; Funduk, N.; Marion, L. The effect of surface grinding and sandblasting on flexural strength and reliability of Y-TZP zirconia ceramic. *Dent. Mater.* **1999**, *15*, 426–433. [CrossRef]
37. Hjerppe, J.; Narhi, T.O.; Vallittu, P.K.; Lassila, L.V. Surface roughness and the flexural and bend strength of zirconia after different surface treatments. *J. Prosthet. Dent.* **2016**, *116*, 577–583. [CrossRef] [PubMed]
38. Park, C.; Vang, M.S.; Park, S.W.; Lim, H.P. Effect of various polishing system on the surface roughness and phase transformation of zirconia and the durability of the polishing systems. *J. Prosthet. Dent.* **2017**, *117*, 430–437. [CrossRef] [PubMed]
39. Al-Wahadni, A.; Martin, D.M. Glasing and Finishing Dental Porcelain. *A Lit. Review. J.-Can. Dent. Assoc.* **1998**, *64*, 580–583.
40. Anusavice, K.J.; Shen, C.; Rawls, H.R. *Phillip's Science of Dental Materials*, 12th ed.; Elsevier: St. Luis, MO, USA, 2012; pp. 231–255.
41. Lai, X.; Si, W.; Jiang, D.; Sun, T.; Shao, L.; Deng, B. Effects of small-grit grinding and glazing on mechanical behaviours and aging resistance of a super-translucent dental zirconia. *J. Dent.* **2017**, *66*, 23–31. [CrossRef]
42. Vasiliu, R.D.; Porojan, S.D.; Bîrdeanu, M.I.; Uțu, I.D.; Porojan, L. The Effect of Thermocycling and Surface Treatments on the Surface Roughness and Microhardness of Three Heat-Pressed Ceramics Systems. *Crystals* **2020**, *10*, 160. [CrossRef]
43. Jum'ah, A.A.; Brunton, P.A.; Li, K.C.; Waddell, J.N. Simulated clinical adjustment and intra-oral polishing of two translucent, monolithic zirconia dental ceramics: An in vitro investigation of surface roughness. *J. Dent.* **2020**, *101*, 103447. [CrossRef]
44. Alao, A.R.; Stoll, R.; Song, X.F.; Miyazaki, T.; Hotta, Y.; Shibata, Y.; Yin, L. Surface quality of yttria-stabilized tetragonal zirconia polycrystal in CAD/CAM milling, sintering, polishing and sand blasting processes. *J. Mech. Behav. Biomed. Mater.* **2017**, *65*, 102–116. [CrossRef]
45. Dib, M.H.M.; Goncalves, A.M.; Jasinevicius, R.G.; Duduch, J.C. Diamond wheel grinding performance evaluation of yttria stabilized zirconia-cubic and tetragonal phases. In Proceedings of the Euspen's 15th International Conference & Exhibition, Leuven, Belgium, 1–5 June 2015.
46. Pereira, G.K.R.; Amaral, M.; Simoneti, R.; Rocha, G.C.; Cesar, P.F.; Valandro, L.F. Effect of grinding with diamond-disc and bur on mechanical behaviour of a Y-TZP ceramic. *J. Mech. Behav. Biomed.* **2014**, *37*, 133–140. [CrossRef]
47. Iseri, U.; Ozkurt, Z.; Yalniz, A.; Kazazoglu, E. Comparison of different grinding procedures on the flexural strange of zirconia. *J. Prosthet. Dent.* **2012**, *107*, 309–315. [CrossRef]
48. Lee, K.; Choe, H.; Heo, Y.; Lee, J.; Son, M. Effect of different grinding burs on the physical properties of zirconia. *J. Adv. Prosthodont.* **2016**, *8*, 137. [CrossRef]
49. Kuraray Noritake—Technical Guide Katana Zirconia Multi-Layered Series. 201805-brochure-katana-zirconia-technical-guide-web-en_2.pdf. Available online: https://www.kuraraynoritake.eu/ (accessed on 21 February 2022).
50. Amann Girrbach-Verarbaitungstehnik—Zolid DNA Generation. 976109DE, 03/2021. Indikationsleitfaden_Zirkon_DE.pdf. Available online: amanngirrbach.com (accessed on 25 January 2022).
51. Ivoclar-Vivadent IPS e.max ZirCAD Labside Gebrauchsinformation 2019-06-27, Rev3a. Available online: https://www.cadstar.dental/download/87/materialdokumentation-de/5667919/ipse-maxzircadlabside.pdf (accessed on 18 February 2022).
52. Pereira, G.K.R.; Silvestri, T.; Camago, R.; Rippe, M.P.; Amaral, M.; Klevarlaan, C.J.; Valandro, L.F. Mechanical behaviour of a Y-TZP ceramic for monolithic restauration: Effect of grinding and low-temperature aging. *Mater. Sci. Eng.* **2016**, *63*, 70–77. [CrossRef] [PubMed]
53. Vasiliu, R.D.; Uțu, I.D.; Porojan, L. The effect of thermocycling on microhardness and surface roughness of two zirconia reinforced lithium silicate glass-ceramics. *J. Matpr.* **2020**, *45*, 4247–4249.
54. Cotes, C.; Arata, A.; Melo, R.M.; Bottino, M.A.; Machado, J.P.B.; Souza, R.O.A. Effects of aging procedure of the topographic surface, structural stability and mechanical strength of ZrO_2-based dental ceramic. *Dent. Mater.* **2014**, *30*, 396–404. [CrossRef] [PubMed]
55. Baldi, A.; Comba, A.; Ferrero, G.; Italia, E.; Tempesta, R.M.; Paolone, G.; Mazzoni, A.; Breschi, L.; Scotti, N. External gap progression after cycling fatigue of adhesive overlays and crowns made with high translucency zirconia or lithium silicate. *J. Eshet. Restor. Dent.* **2022**, *34*, 557–564. [CrossRef] [PubMed]
56. Reyes, A.R.; Dennisson, J.B.; Powers, J.M.; Sierraalta, M.; Yaman, P. Translucency and Flexural Strength of Translucent Zirconia Ceramics. *J. Prosthet. Dent.* 2021, *in press*.

57. Jones, C.S.; Billington, R.W.; Pearson, G.J. The in vivo perception of roughness of restauration. *Br. Dent. J.* **2004**, *196*, 42–45. [CrossRef]
58. Bollen, C.M.; Lambrechts, P.; Quirynen, M. Comparison of surface roughness of oral hard materials to the threshold surface roughness for bacterial plaque retention: A review of the literature. *Dent. Mater. J.* **1997**, *13*, 258–269. [CrossRef]

materials

MDPI

Article

Evaluation of Physical Properties of Zirconia Suspension with Added Silane Coupling Agent for Additive Manufacturing Processes

Jae-Gon Jang [1,†], Jin-Ho Kang [1,†], Kwang-Bum Joe [2], Kumaresan Sakthiabirami [1,3], Kyoung-Jun Jang [1], Mee-Jin Jun [4], Gye-Jeong Oh [3], Chan Park [1] and Sang-Won Park [1,3,*]

1 Department of Prosthodontics, School of Dentistry, Chonnam National University, Gwangju 61186, Korea; jangjaegon@naver.com (J.-G.J.); jhk.bme1002@gmail.com (J.-H.K.); sakthikarthi.dentist@gmail.com (K.S.); zircong@naver.com (K.-J.J.); upgradepc@hanmail.net (C.P.)
2 School of Dentistry, Chonnam National University, Gwangju 61186, Korea; poopoopooy@naver.com
3 Biomedical Evaluation and Research Centre, School of Dentistry, Chonnam National University, Gwangju 61186, Korea; timer22@naver.com
4 Department of Dental Hygiene, Gwangju Health University, Gwangju 62287, Korea; jmejin@hanmail.net
* Correspondence: psw320@chonnam.ac.kr; Tel.: +82-62-530-5842
† These authors contributed equally to this work.

Abstract: In this study, we have analysed the effects of a silane coupling agent on the volume fraction of zirconia for digital light processing (DLP)-based additive manufacturing processes. Zirconia suspension was prepared by the incorporation of silane-modified zirconia particles (experimental group) or untreated zirconia particles (control group). Furthermore, the control and experimental group were subdivided into three groups based on the volume fraction (52, 54, and 56 vol%) of zirconia particles. The disk-shaped zirconia samples were 3D (three-dimensional) printed using the DLP technique and their physical and mechanical properties were evaluated. The addition of a silane coupling agent to the zirconia samples was found to have influence of about 6% on the hardness and biaxial flexural strength. Moreover, the decrease in minute air gaps inside the zirconia layers significantly increased the material density (visualized from the microstructure analysis). Thus, from this study, it was established that the silane-modified zirconia particles had a positive effect on the physical properties of the zirconia parts.

Keywords: additive manufacturing; digital light processing (DLP) technology; zirconia suspension; silane coupling agent; physical property

Citation: Jang, J.-G.; Kang, J.-H.; Joe, K.-B.; Sakthiabirami, K.; Jang, K.-J.; Jun, M.-J.; Oh, G.-J.; Park, C.; Park, S.-W. Evaluation of Physical Properties of Zirconia Suspension with Added Silane Coupling Agent for Additive Manufacturing Processes. *Materials* **2022**, *15*, 1337. https://doi.org/10.3390/ma15041337

Academic Editors: Rodrigo Moreno and Matthias Kern

Received: 24 December 2021
Accepted: 9 February 2022
Published: 11 February 2022

1. Introduction

Various materials have been used for prostheses in dental clinics, such as titanium, cobalt–chromium, zirconia, and glass ceramics [1–7]. Among them, ceramic has been widely used and actively investigated, especially with the development of computer-aided design/computer-aided manufacturing (CAD/CAM) technology. Thus far, dental ceramics that have been widely utilised in clinics include dental Empress and InCeram Alumina. However, these materials do not satisfy the physical properties, such as strength, required by dental clinics [1–3]. To improve the properties of these ceramics, research into zirconia is ongoing. Several studies on zirconia have found that this compound enhances material fracture strength and biocompatibility; it has been widely applied as a material for dental clinic-fabricated prostheses and artificial joints [8]. Zirconia ceramic prostheses have been increasingly used because patients demand better mechanical properties and aesthetics [9,10]. In addition, CAD/CAM technology has been applied to clinic care for dental restorations employing zirconia. However, the use of CAD/CAM to produce zirconia prostheses may produce rough scratches or form defects on the prosthesis surface, eventually leading to cracks and fracture. Furthermore, the fabrication of highly complex shapes using CAD/CAM is extremely difficult [10–12]. To overcome these disadvantages, research on additive manufacturing and formulations for additive manufacturing is actively being conducted. Additive manufacturing is a method that involves stacking the layers of materials using digital image processing techniques [13]. Recent research indicates that dental prostheses manufactured through additive manufacturing present acceptable density compared with those produced by milling or traditional methods [14–18].

Additive manufacturing includes spot scanning-based stereolithography, which uses a stereolithography apparatus (SLA), digital light processing (DLP), and Polyjet printing based on mask image projection [7]. Of these, SLA and DLP are technologies that print in layers using a photocuring reaction of photosensitive resins. The DLP technology stacks and hardens a formulated liquid photocurable resin through a photocuring process involving projected light according to the type and location of the shape to be moulded [19,20]. The main advantages of DLP are as follows: the printed product is highly dense and has a luminous surface, and the work speed is fast because the resin is moulded in face units. In addition, DLP has a faster production speed than the SLA process and enables the low-cost fabrication of samples with a small sample size [20,21]. For these reasons, the DLP system is considered suitable for additive manufacturing printers for dental ceramic production. However, despite the advances in this technology, preventing the formation of micro-cracks on the zirconia surface is difficult. The physical properties (e.g., strength) of the product produced by additive manufacturing are found to be extremely inadequate for actual clinical application [22]. To resolve this issue, a silane coupling agent can be added to the suspension for additive manufacturing. The agent combines with the surface of inorganic fillers or glass fibres that are mixed with the polymer on one side of the molecules, but has an affinity with the matrix resin on the other side. Dispersibility is enhanced, and the interfacial adhesiveness among inorganic fillers or between the glass fibre and matrix polymer improves. An organic coating based on a silane coupling agent is applied to enhance surface characteristics [23–26]. Its use was found to improve the strength of each sample. Studies indicate that the silane coupling agent improves the zirconia surface by promoting properties, such as surface strength and abrasion as well as friction resistance [27]. More importantly, silane coupling agents reduce the number of hydroxyl groups on the surface of the modified filler, increasing hydrophobicity and improving dispersion in polymer matrices and resins. This has been found to improve the strength of samples by limiting the hydrophilicity of the filler and increasing the reactivity of the filler or compatibility with the selected polymer depending on the type of silane functional group [28–31].

Methyltrimethoxysilane (MTMS), an organosilane with one methyl group and three hydrolyzable methoxy substituents, has been successfully used to impart hydrophobicity and oleophobicity to a variety of substrates. As previously reported, MTMS controls the hydrolysis conditions, so that the hydrolysis and coating processes are usually in simple ambient aqueous conditions, which represents an obvious advantage and is easily compatible with existing manufacturing processes [32].

In this study, the zirconia suspensions with a volume fraction of 52–56% without added silane were set as the control group [33], whereas those with the coupling agent were set as the experimental groups. The groups were prepared to compare the physical properties of zirconia specimens after sintering based on the incorporation or absence of a silane coupling agent. The null hypothesis of this study is that there is no difference in the physical and mechanical properties of the 3D printed zirconia samples with and without the addition of a silane coupling agent according to volume fraction.

2. Materials and Methods

2.1. Preparation of Zirconia Photocurable Suspension

Commercially available zirconia powder (TZ-3Y, Tosho, Japan) with an average particle size of 310 nm was used. Three monomers (all obtained from Sigma Aldrich Inc., St. Louis, MO, USA), i.e., acrylic resin-based IBA (isobornyl acrylate), HDDA (1,6-hexanediol diacrylate), and PNPGDA (propoxylated neopentyl glycol diacrylate), were utilised to manufacture a photocurable binder. Other additives were a photoinitiator (Irgacure 819, Ciba Specialty Chemicals, Swiss), a dispersant to improve the dispersibility of zirconia powder binder (BYK-180, BYK Inc., Wesel, Germany), and a silane coupling agent (methyltrimethoxysilane coupling agent, MTMS, Duksan, Korea). MTMS mixed solution was prepared by stirring at a mass ratio of ethanol, distilled water, and MTMS (90:8:2) for 1 h. Then, 100 g of zirconia was mixed with the MTMS mixed solution. After adding the particles, it was ball milled for 24 h and then dried overnight at room temperature. The physical properties of the silane-modified zirconia particles were evaluated using FT-IR spectroscopy (Spectrum 400, Perkin Elmer, USA). The silane-modified zirconia powders were used to prepare suspensions of zirconia volume fractions of 52, 54, and 56 vol%. A planetary centrifugal mixer (ARV-310, Thinky Corp., Tokyo, Japan) was used to mix the suspension homogeneously. The groups were mainly divided into Z (control group) and ZS (experimental group with silane coupling agent); further, it was sub-divided into three control groups and three experimental groups based on zirconia volume fraction. The zirconia suspension compositions are summarised in Table 1. The detailed schematic illustration of the experimental group process was displayed in Figure 1.

Table 1. Experimental groups in this study.

Group		Zirconia Powder	Acrylic Monomers			Additive Agents		Total
			IBA	HDDA	PNPGDA	Dispersant	Silane Coupling Agent	wt (vol)%
Control	Z52	86.45 (52.00)		9.23 (33.43)		4.32(14.57)	0	100
	Z54	87.34 (54.00)		8.29 (30.87)		4.37(15.13)	0	100
	Z56	88.19 (56.00)		7.41 (28.32)		4.41(15.68)	0	100
Experimental	ZS52	86.45 (52.00)		8.64 (30.30)		3.27(11.65)	1.64 (6.05)	100
	ZS54	87.34 (54.00)		7.68 (27.62)		3.32(12.10)	1.66 (6.28)	100
	ZS56	88.19 (56.00)		6.74 (24.94)		3.38(12.54)	1.69 (6.52)	100

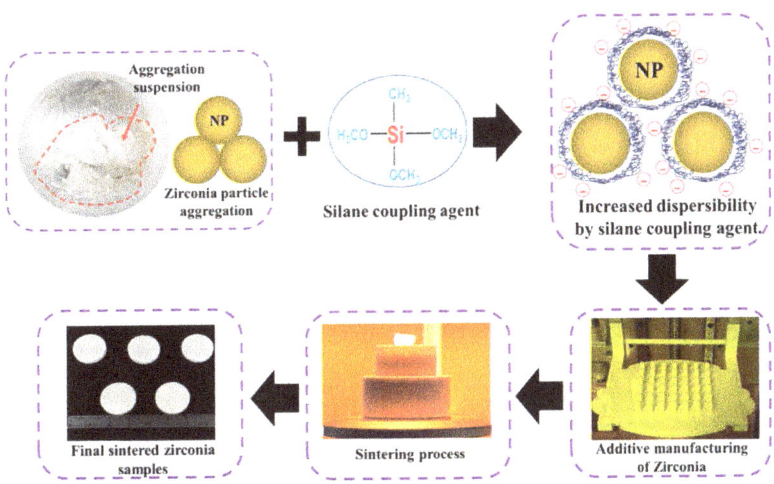

Figure 1. Schematic illustration of the experimental group process.

2.2. Zirconia Samples Fabrication Using Additive Manufacturing

A DLP machine was used for light-curing the zirconia suspension (Octave Light R1, Octave Light Ltd., Shatin, Hongkong) (Figure 2); the waveband was 385 nm. Zirconia samples were additively manufactured by setting the thickness of each layer to 50 μm. The STL files of samples were designed using a three-dimensional (3D) modelling program (Solid Works 2016, Dassault Systemes SOLIDWORKS Corp., Waltham, MA, USA). Each sample was fabricated in the form of a disk with diameter and height of 20 and 2 mm, respectively.

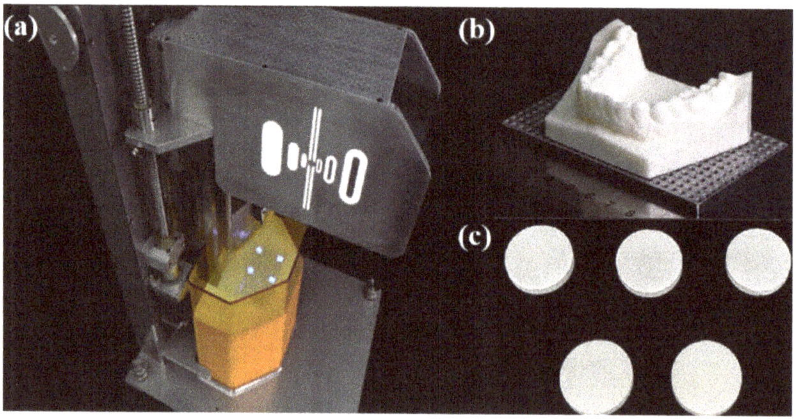

Figure 2. DLP machine used for preparing zirconia samples: (**a**) DLP machine curing formulations; (**b**) mandible and teeth model manufactured using DLP machine; (**c**) zirconia disk samples manufactured by DLP machine.

2.3. Degreasing and Sintering of Zirconia Samples

The additively manufactured zirconia samples were degreased and sintered using a sintering machine (DUOTRON PRO ex-6100, Zirconia Sintering Furnace, ADDIN Co., LTD, Suwon, Korea). The degreasing of each experimental group proceeded as follows. The temperature was gradually increased by 0.5 °C/min until it reached 200 °C and maintained for 1 h. Then, it was increased to 300 °C at 0.5 °C/min and again held for 1 h. Finally, the temperature was incrementally increased to 500 °C at 0.5 °C/min and maintained for 1 h. After the degreasing process, the sintering process proceeded as follows. The temperature was gradually increased to 1450 °C at 10 °C/min and held for 2 h. Then, the samples were naturally cooled (Figure 3) [26]. The final sintered zirconia samples were 14.5 and 2 mm in diameter and height, respectively.

2.4. Linear Shrinkage

The linear shrinkage of fabricated zirconia samples (n = 10) was measured using a vernier calliper (ASTM C 326: 2009). The diameter and height of samples were measured before and after sintering according to the volume fraction of zirconia. The following formula is used:

$$\triangle X = \{(X_1 - X_2)/X_1\} \times 100, \tag{1}$$

where X_1 and X_2 are the volume (mm^3) of zirconia before and after sintering, respectively.

2.5. Relative Density Measurement

The relative density of zirconia samples was measured according to the volume fraction with reference to ISO 18754: 2007. After grinding the surface to reduce the influence of surface roughness, relative density is calculated using the following formula:

$$\rho = (m_1/m_3 - m_2) \times \rho_{(Liq)}, \tag{2}$$

where m_1 is the dry mass (g), m_2 is the underwater mass (g), m_3 is water mass (g), and $\rho_{(Liq)}$ is the density of liquid infiltration solution (density of water at room temperature = 1.0 g/m^3); relative density (%) is calculated by (measured density/theoretical density) × 100.

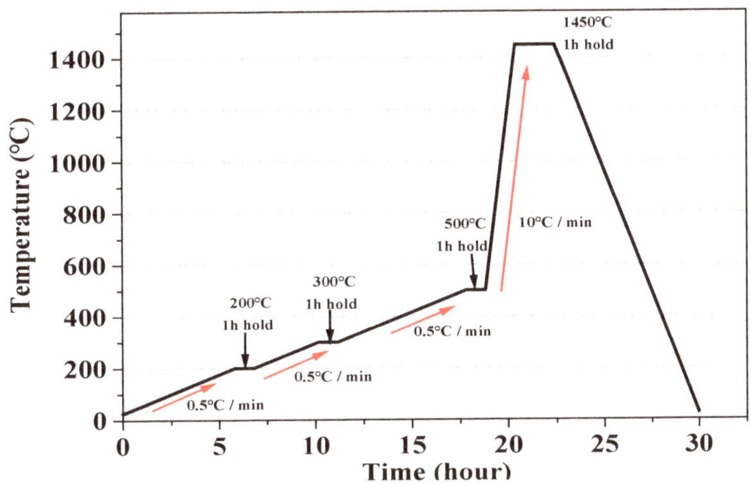

Figure 3. Sintering process of zirconia specimens (The red arrow indicates the heating rate.).

2.6. Hardness

To measure the hardness of the final sintered zirconia samples (each with a diameter and height of 14.5 and 2 mm, respectively), these were fixed to multiple diamond plates and ground under constant pressure using grits in the order #400, #800, #1200, and #2400 (n = 10). The sample surface was finely ground using diamond pastes with 3 μm and 1 μm particles (MP-DS-50-0300, MTDI, Daejeon, Korea). The ground samples were pressed five times for 12 s with a force of 1 kgf (9.8 N) using a Micro-Vickers hardness tester (NOVA® 130/240, INNOVATEST, Maastricht, Nederland) to determine the micro-hardness at each instance.

2.7. Biaxial Flexural Strength

The flexural strength of zirconia samples pertains to the bending stress that the samples can withstand when pressure is applied to them. The biaxial flexural strength of samples depending on the volume fraction was measured according to ISO 6872. Ten samples from each experimental group were measured by loading in a universal testing machine (RB Model 301 Unitech MTM R&B, Seongnam, Korea). The load was increased at a cross-head speed of 1 mm/min until the samples were completely fractured. Calculations are performed using the following formulae:

$$\sigma = -0.2387P (X - Y)/b2, \quad (3)$$

$$X = (1 + v) \ln(r2/r3)2 + [(1 - v)/2] (r2/r3) (2), \quad (4)$$

$$Y = (1 + v) [1 + \ln(r1/r3)2] + (1 - v) (r1/r3) (2), \quad (5)$$

where σ is the biaxial flexural strength (MPa); P is the maximum load at fracture (Newtons); b is the sample thickness (mm); v is Poisson's Ratio (v = 0.25); r1 is the radius of the circle supporting the load (mm); r2 is the radius of the circle to which the load is applied (mm); and r3 is the sample radius (mm).

2.8. Evaluation of Microstructure of Final Sintered Zirconia Samples

To observe the microstructure of zirconia samples, their surfaces were ground using diamond pastes with 3 μm and 1 μm particles (MP-DS-50-0300, MTDI, Daejeon, Korea). Then, they were cleaned with alcohol and distilled water in an ultrasonic cleaner (JAC-2010, KODO Technical Research Co., Hwaseong, Korea) for 20 min. To observe the cross-section, the samples were cleaned with alcohol and distilled water for 20 min. The platinum coating was applied to sample surfaces using a sputter coater (108 Auto, Cressington Scientific Instruments, Ltd., Oxhey, UK). After coating, the zirconia sample layers were observed using a scanning electron microscope (FE-SEM, JEOL, JSM-7500F, Tokyo, Japan).

2.9. Statistical Analysis

The sample size was evaluated using G-Power 3.1 software (University of Dusseldorf, Dusseldorf, Germany) [5]. The experimental results were statistically processed by one-way analysis of variance using the SPSS software package (SPSS Version 20.0, SPSS Inc., Chicago, IL, USA), and post tested with the Tukey HSD test. All results were tested for statistical significance with $p < 0.05$.

3. Results

3.1. FT-IR Analysis

FT-IR spectroscopy was performed to determine the organic characterization, i.e., the effect of silanization of zirconia particles. As shown in Figure 4, the stretching vibration of -CH appears at 2983 cm^{-1} and 2900 cm^{-1}, respectively. The bending vibration of –CH is 1405 cm^{-1}. At 1270 cm^{-1} the stretching vibration of C–O appears, while the stretching vibration of –Si–O– attached to the –CH is at 1052 cm^{-1}, the stretching vibration of –O–O– appears at 890 cm^{-1}. From the results, it was confirmed that the zirconia particles were treated with the silane coupling agent, and a zirconia slurry was prepared using the treated powder.

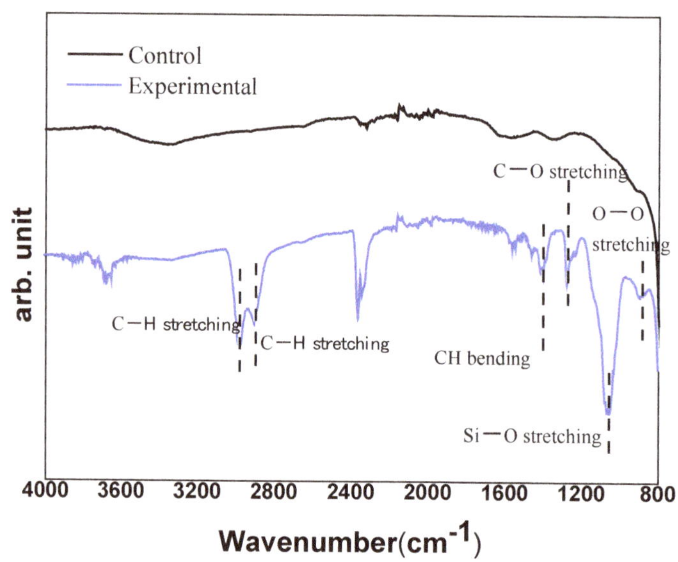

Figure 4. FT-IR spectra of the zirconia powder before and after silanization.

3.2. Linear Shrinkage

The linear shrinkage of additively manufactured zirconia samples was measured according to the zirconia volume fraction by comparing them before and after sintering (Table 2 and Figure 5).

Table 2. Average linear shrinkage of printed zirconia sample.

Groups	Linear Shrinkage (%)	
	Control	Experimental
Z52	19.20 ± 0.31	19.14 ± 0.17
Z54	19.31 ± 0.23	19.23 ± 0.20
Z56	19.15 ± 0.29	19.21 ± 0.09

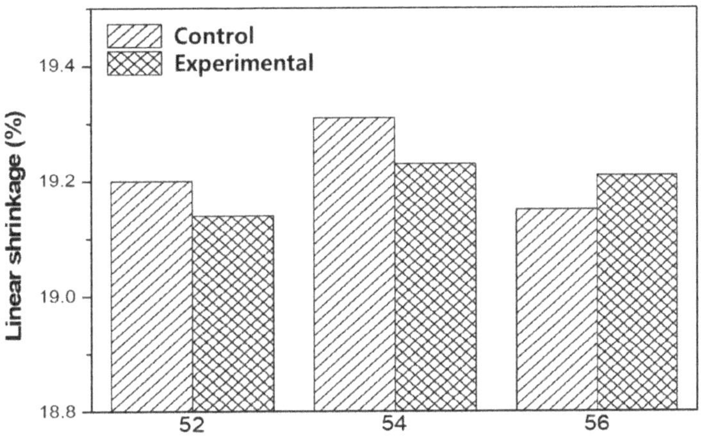

Figure 5. Comparison of linear shrinkage between control and experimental groups after final sintering.

In the previous experiment, the shrinkage range of zirconia powder samples sintered at 1200–1400 °C was 19–20%; consequently, the average linear shrinkage of zirconia ceramic was found to be 18–25% [11]. The linear contraction rate of experimental and control groups exhibited a constant pattern of approximately 18–20%; however, no significant difference in the contraction rate between the two groups was observed. Moreover, it should be noted that in all groups, linear shrinkage was greater in height than in diameter. This is attributed to the occurrence of linear shrinkage in sample height during sintering, indicating that the action of gravitational force on height exceeds the effect of this force on diameter. Moreover, the greater linear shrinkage in height than in diameter was presumed to occur because the evaporation of resin in each layer at high temperatures caused the shrinkage of air gaps [34,35].

3.3. Relative Density

The relative density values of additively manufactured zirconia samples in control groups and experimental groups are shown in Table 3 and Figure 6.

Table 3. Relative density of printed zirconia; different letters represent significant differences ($p < 0.05$).

Groups	Relative Density (%)	
	Control	Experimental
Z52	94.89 ± 0.35 [a]	19.14 ± 0.09 [a]
Z54	95.65 ± 0.67 [b]	19.06 ± 0.10 [b]
Z56	96.15 ± 0.59 [c]	19.05 ± 0.03 [d]

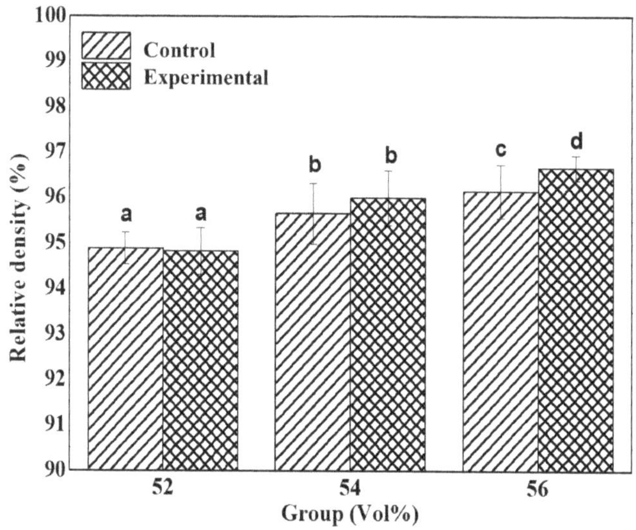

Figure 6. Comparison of relative density between control and experimental groups; different letters represent significant differences ($p < 0.05$).

The difference between Z52, Z54, and Z56 was significant ($p < 0.05$). As for the experimental groups, the difference between ZS54 and ZS56 was insignificant, whereas that between these two and ZS52 was significant. Based on the addition of silane coupling agent, the difference between Z52 and ZS52 and that between Z54 and ZS54 did not significantly differ; however, a significant difference between Z56 and ZS56 ($p > 0.05$) was observed. Regardless of whether the silane coupling agent was added to the samples, the relative density increased as the volume fraction of zirconia increased. However, when a high-volume fraction of zirconia was added, density was further increased with the addition of a silane coupling agent. The silane coupling agent was presumed to enable the homogeneous dispersion of zirconia particles in the suspension. As a result, density was evenly distributed during particle growth [36]. The higher the relative density, the higher the sample density; hence, mechanical properties, such as strength, can be enhanced.

3.4. Hardness Measurement

The Vickers hardness evaluation results of control and experimental groups were shown in Table 4 and Figure 7.

Table 4. Vickers hardness of printed zirconia; different letters represent significant differences ($p < 0.05$).

Groups	Vickers hardness (HV)	
	Control	Experimental
Z52	1573 ± 21.31 ^{a}	1666 ± 14.61 ^{d}
Z54	1648 ± 48.17 ^{b}	1748 ± 20.33 ^{e}
Z56	1777 ± 22.19 ^{c}	1848 ± 24.04 ^{f}

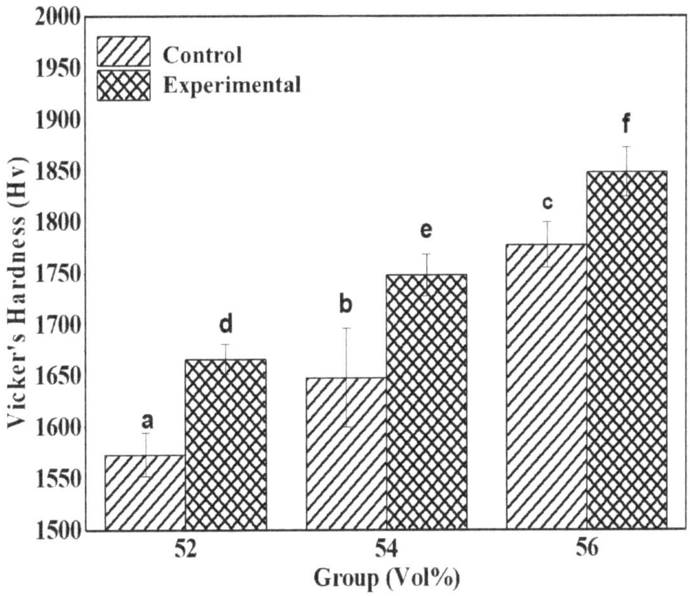

Figure 7. Comparison of Vickers hardness between control and experimental groups; different letters represent significant differences ($p < 0.05$).

All experimental groups significantly differed in average hardness ($p < 0.05$). All zirconia samples exhibited significant variations in Vickers average hardness ($p < 0.05$). Group ZS56, which had a zirconia volume fraction of 56 vol% and contained a silane coupling agent, exhibited the highest hardness value; it was followed by Z56, ZS54, ZS52, Z54, and Z52. The hardness values among these groups significantly differed (Figure 5). The addition of silane coupling agents to the samples promoted the homogeneous commixture of organic (photocurable polymers) and inorganic (zirconia) substances. As a result, dispersibility improved and hardness increased [21,28,36].

3.5. Biaxial Flexural Strength

From the biaxial flexural strength measurement of samples, the average biaxial flexural strength was displayed in Table 5 and Figure 8.

Table 5. Vickers hardness of printed zirconia; different letters represent significant differences ($p < 0.05$).

Group	Flexural Strength (MPa)	
	Control	Experimental
Z52	346 ± 6.96 [a]	368 ± 8.62 [d]
Z54	384 ± 3.68 [b]	407 ± 8.87 [e]
Z56	407 ± 8.73 [c]	433 ± 3.27 [f]

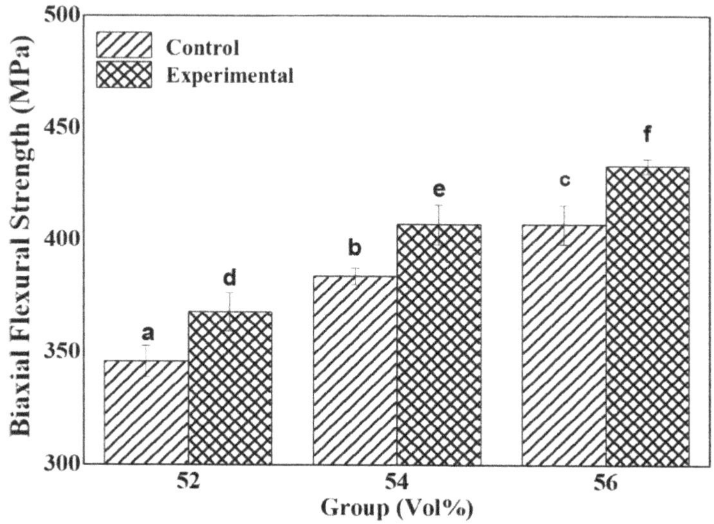

Figure 8. Comparison of biaxial flexural strength between control and experimental groups; different letters represent significant differences ($p < 0.05$).

The biaxial flexural strengths between Z52 and ZS52, between Z54 and ZS54, and between Z56 and ZS56 significantly differed according to the added silane coupling agent ($p < 0.05$). Similar to the results of Vickers hardness measurement, the following results are attributed to the commixture of photocuring polymer and zirconia promoted by the addition of a silane coupling agent, thereby increasing dispersibility [37]. Accordingly, the biaxial flexural strength of groups with the silane coupling agent increased.

3.6. Microstructure Evaluation

The fractured sections of additively manufactured zirconia samples were observed using the scanning electron microscope for microstructure evaluation according to the zirconia volume fraction (Figure 9). The layers of additively manufactured zirconia samples were examined through this observation. The smaller the zirconia volume fraction, the more air gaps were observed (red arrows in Figure 9). The variations in relative density measurements were confirmed to be caused by these air gaps.

These observations are considered to support the measurements of low density and strength because of the occurrence of numerous air gaps in groups with no added silane coupling agent. In addition, the air gaps in groups with the added silane coupling agent were smaller in size compared with those in groups with no added coupling agent.

The images of the layers of additively manufactured zirconia samples were captured through scanning electron microscopy. The layers were observed to be smaller and more constant in groups with the added silane coupling agent compared with those without the agent. Moreover, the proportion of air gaps caused by the debinding of zirconia samples during sintering was found to decrease and then remain constant. If the air gaps were not constant, micro-cracks were more likely to occur due to external forces that may degrade strength, hardness, and physical properties [33,38]. If zirconia-based prostheses are applied in actual dental clinical practice, the foregoing can increase the possibility of prosthesis breakage caused by the masticatory force of patients.

Therefore, considering that the air gaps are reduced and remain constant, improved strength can be expected from additively manufactured zirconia with the added silane coupling agent.

This study demonstrated the effect of silane on the volume fraction of zirconia. However, future experiments are needed to evaluate the dispersion of the silane-modified zirconia particles in the suspension by varying experimental parameters associated with the salinisation of zirconia particles, such as pH, solution composition, and hydrolysis time.

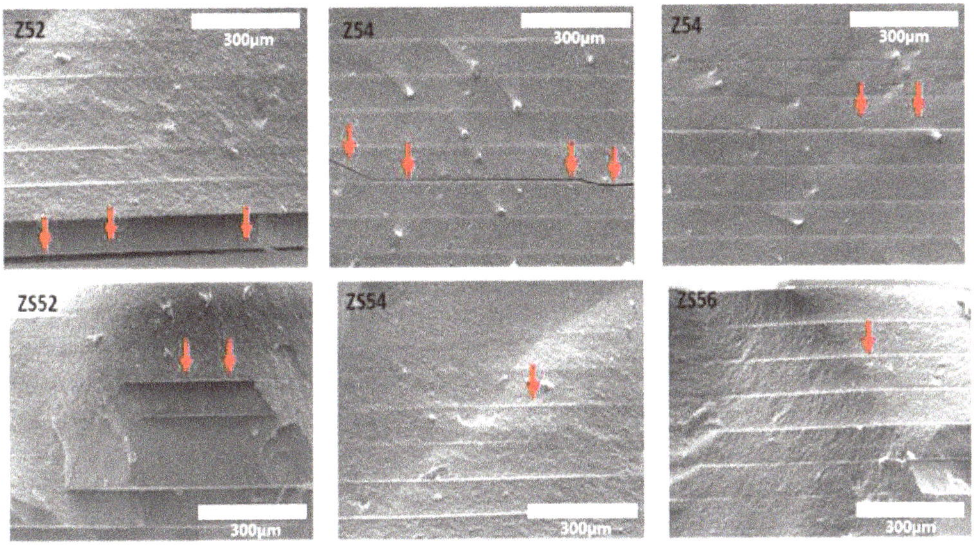

Figure 9. FE-SEM images of microstructure and layer-by-layer observation of the final sintered zirconia specimen (Red arrows indicate microcracks and pores).

4. Conclusions

In this study, the preparation of zirconia suspensions with the addition of a silane coupling agent (approximately 2% of zirconia volume fraction) for DLP additive manufacturing technology has been demonstrated. It was established that the density of the printed zirconia samples significantly increased with an increase in the zirconia volume fraction (52, 54, and 56 vol%). Interestingly, the experimental group (ZS56) with the addition of silane coupling agent to the suspension of 56 vol% exhibits improved strength and hardness (5–6%) compared to those without silane coupling agent (control group). However, it is essential to focus future research activities on enhancing the dispersibility of zirconia suspension using the silane coupling agent for DLP printing. In future experiments, we plan to evaluate the dispersion stability of the zirconia suspension by varying conditions, such as pH control, solution composition, and hydrolysis time, for the encapsulation of the zirconia particles with the silane coupling agent.

Author Contributions: Conceptualization and writing—original draft preparation, S.-W.P., J.-G.J. and J.-H.K.; writing—review and editing, S.-W.P., J.-H.K., K.S. and C.P.; Methodology, K.-B.J. and J.-G.J.; visualization, K.-J.J. and M.-J.J.; supervision and funding acquisition, G.-J.O. and S.-W.P. All authors have read and agreed to the published version of the manuscript.

Funding: This work was supported by a National Research Foundation of Korea Grant funded by the Korean government (MSIP) [Grant no. 2019R1A2C108945613 and 2020R111A1A01073015].

Institutional Review Board Statement: Not applicable.

Informed Consent Statement: Not applicable.

Data Availability Statement: All the data have been illustrated in the manuscript.

Conflicts of Interest: The authors declare no conflict of interest.

References

1. Mclean, J.W. The reinforcement of dental porcelain with ceramic oxides. *Br. Dent. J.* **1965**, *119*, 251–267. [PubMed]
2. Shareef, M.; Van Noort, R.; Messer, P.; Piddock, V. The effect of microstructural features on the biaxial flexural strength of leucite reinforced glass-ceramics. *J. Mater. Sci. Mater. Med.* **1994**, *5*, 113–118. [CrossRef]
3. Jones, D.W. Development of dental ceramics. An historical perspective. *Dent. Clin. North Am.* **1985**, *29*, 621–644. [PubMed]
4. Ammarullah, M.I.; Afif, I.Y.; Maula, M.I.; Winarni, T.I.; Tauviqirrahman, M.; Akbar, I.; Basri, H.; van der Heide, E.; Jamari, J. Tresca Stress Simulation of Metal-on-Metal Total Hip Arthroplasty during Normal Walking Activity. *Materials* **2021**, *14*, 7554. [CrossRef] [PubMed]
5. Ramaraj, S.; Kim, M.-A.; Rosa, V.; Prasanna, N.; Shon, W.-J.; Min, K.-S. Combined Effect of Melittin and DNase on Enterococcus faecalis Biofilms and Its Susceptibility to Sodium Hypochlorite. *Materials* **2020**, *13*, 3740. [CrossRef] [PubMed]
6. Sakthiabirami, K.; Kang, J.H.; Jang, J.G.; Lim, H.P.; Yun, K.D.; Park, C.; Yim, E.K.; Park, S.W.; Ko, K.T.; Oh, G.J. Effect of Argon Plasma Treatment on Surface-Treated Titanium. *Adv. Dent. Oral Health* **2019**, *11*, 158–164.
7. Kumaresan, S.; Vaiyapuri, S.; Kang, J.-H.; Dubey, N.; Manivasagam, G.; Yun, K.-D.; Park, S.-W. Additive Manufactured Zirconia-Based Bio-Ceramics for Biomedical Applications [Online First], IntechOpen. 19 January 2022. Available online: https://www.intechopen.com/online-first/80125 (accessed on 10 February 2022). [CrossRef]
8. Sakthiabirami, K.; Kim, J.W.; Kang, J.H.; Jang, K.J.; Oh, G.J.; Fisher, J.G.; Yun, K.D.; Lim, H.P.; Park, S.W. Tailoring interfacial interaction through glass fusion in glass/zinc-hydroxyapatite composite coatings on glass-infiltrated zirconia. *Ceram. Int.* **2018**, *44*, 16181–16190. [CrossRef]
9. Schriwer, C.; Skjold, A.; Gjerdet, N.R.; Øilo, M. Monolithic zirconia dental crowns. Internal fit, margin quality, fracture mode and load at fracture. *Dent. Mater.* **2017**, *33*, 1012–1020. [CrossRef]
10. D'Addazio, G.; Santilli, M.; Rollo, M.L.; Cardelli, P.; Rexhepi, I.; Murmura, G.; Al-Haj Husain, N.; Sinjari, B.; Traini, T.; Özcan, M. Fracture resistance of zirconia-reinforced lithium silicate ceramic crowns cemented with conventional or adhe-sive systems: An in vitro study. *Materials* **2020**, *13*, 2012. [CrossRef]
11. Denry, I.; Kelly, J.R. State of the art of zirconia for dental applications. *Dent. Mater.* **2008**, *24*, 299–307. [CrossRef]
12. Preis, V.; Behr, M.; Hahnel, S.; Handel, G.; Rosentritt, M. In vitro failure and fracture resistance of veneered and full-contour zirconia restorations. *J. Dent.* **2012**, *40*, 921–928. [CrossRef] [PubMed]
13. Preis, V.; Behr, M.; Handel, G.; Schneider-Feyrer, S.; Hahnel, S.; Rosentritt, M. Wear performance of dental ceramics after grinding and polishing treatments. *J. Mech. Behav. Biomed. Mater.* **2012**, *10*, 13–22. [CrossRef] [PubMed]
14. Huang, Z.; Zhang, L.; Zhu, J.; Zhao, Y.; Zhang, X. Clinical marginal and internal fit of crowns fabricated using different CAD/CAM technologies. *J. Prosthodont.* **2015**, *24*, 291–295. [CrossRef] [PubMed]
15. Dawood, A.; Marti, B.M.; Sauret-Jackson, V.; Darwood, A. 3D printing in dentistry. *Br. Dent. J.* **2015**, *219*, 521–529. [CrossRef]
16. Park, J.-M.; Ahn, J.-S.; Cha, H.-S.; Lee, J.-H. Wear resistance of 3D printing resin material opposing zirconia and metal antagonists. *Materials* **2018**, *11*, 1043. [CrossRef] [PubMed]
17. Zeltser, C.; Lewinstein, I.; Grajower, R. Fit of crown wax patterns after removal from the die. *J. Prosthet. Dent.* **1985**, *53*, 344–346. [CrossRef]
18. Abduo, J.; Lyons, K.; Swain, M. Fit of zirconia fixed partial denture: A systematic review. *J. Oral Rehabil.* **2010**, *37*, 866–876. [CrossRef]
19. Harish, V.; Mohamed Ali, S.; Jagadesan, N.; Mohamed Ifthikar, S.S.; Debasish Basak, F.H. Evaluation of internal and marginal fit of two metal ceramic system—In vitro study. *J. Clin. Diagn. Res. JCDR* **2014**, *8*, ZC53.
20. He, R.; Liu, W.; Wu, Z.; An, D.; Huang, M.; Wu, H.; Jiang, Q.; Ji, X.; Wu, S.; Xie, Z. Fabrication of complex-shaped zirconia ceramic parts via a DLP-stereolithography-based 3D printing method. *Ceram. Int.* **2018**, *44*, 3412–3416. [CrossRef]
21. Kang, J.-H.; Jang, K.-J.; Sakthiabirami, K.; Oh, G.-J.; Jang, J.-G.; Park, C.; Lim, H.-P.; Yun, K.-D.; Park, S.-W. Mechanical properties and optical evaluation of scaffolds produced from 45S5 bioactive glass suspensions via stereolithography. *Ceram. Int.* **2020**, *46*, 2481–2488. [CrossRef]
22. Sakthiabirami, K.; Soundharrajan, V.; Kang, J.-H.; Yang, Y.P.; Park, S.-W. Three-Dimensional Zirconia-Based Scaffolds for Load-Bearing Bone-Regeneration Applications: Prospects and Challenges. *Materials* **2021**, *14*, 3207. [CrossRef] [PubMed]
23. Moreno, M.B.P.; Murillo-Gómez, F.; de Goes, M.F. Physicochemical and morphological characterization of a glass ceramic treated with different ceramic primers and post-silanization protocols. *Dent. Mater.* **2019**, *35*, 1073–1081. [CrossRef] [PubMed]
24. Moreno, M.B.P.; Murillo-Gómez, F.; De Goes, M. Effect of different ceramic-primers and silanization-protocols on glass-ceramic bond strength. *Dent. Mater.* **2018**, *34*, e80–e81. [CrossRef]
25. Villa, S.; Riani, P.; Locardi, F.; Canepa, F. Functionalization of Fe3O4 NPs by silanization: Use of amine (APTES) and thiol (MPTMS) silanes and their physical characterization. *Materials* **2016**, *9*, 826. [CrossRef]
26. Jang, K.-J.; Kang, J.-H.; Fisher, J.G.; Park, S.-W. Effect of the volume fraction of zirconia suspensions on the microstructure and physical properties of products produced by additive manufacturing. *Dent. Mater.* **2019**, *35*, e97–e106. [CrossRef]
27. Lopez de Armentia, S.; Pantoja, M.; Abenojar, J.; Martinez, M.A. Development of silane-based coatings with zirconia nanoparticles combining wetting, tribological, and aesthetical properties. *Coatings* **2018**, *8*, 368. [CrossRef]
28. Barczewski, M.; Matykiewicz, D.; Szostak, M. The effect of two-step surface treatment by hydrogen peroxide and silani-zation of flax/cotton fabrics on epoxy-based laminates thermomechanical properties and structure. *J. Mater. Res. Technol.* **2020**, *9*, 13813–13824. [CrossRef]

29. Nakonieczny, D.S.; Antonowicz, M.; Paszenda, Z. Surface modification methods of ceramic filler in ceramic-carbon fibre composites for bioengineering applications—A systematic review. *Rev. Adv. Mater. Sci.* **2020**, *59*, 586–605. [CrossRef]
30. Nakonieczny, D.S.; Kern, F.; Dufner, L.; Dubiel, A.; Antonowicz, M.; Matus, K. Effect of Calcination Temperature on the Phase Composition, Morphology, and Thermal Properties of ZrO2 and Al2O3 Modified with APTES (3-aminopropyltriethoxysilane). *Materials* **2021**, *14*, 6651. [CrossRef]
31. Lin, H.; Rosu, C.; Jiang, L.; Sundar, V.A.; Breedveld, V.; Hess, D.W. Nonfluorinated superhydrophobic chemical coatings on polyester fabric prepared with kinetically controlled hydrolyzed methyltrimethoxysilane. *Ind. Eng. Chem. Res.* **2019**, *58*, 15368–15378. [CrossRef]
32. Song, S.Y.; Park, M.S.; Lee, J.W.; Yun, J.S. A study on the rheological and mechanical properties of photo-curable ceramic/polymer composites with different silane coupling agents for SLA 3D printing technology. *Nanomaterials* **2018**, *8*, 93. [CrossRef] [PubMed]
33. Jang, K.-J.; Kang, J.-H.; Sakthiabirami, K.; Lim, H.-P.; Yun, K.-D.; Yim, E.-K.; Oh, G.-J.; Yang, H.-S.; Lee, K.-K.; Park, S.-W. Evaluation of cure depth and geometrical overgrowth depending on zirconia volume fraction using digital light processing. *J. Nanosci. Nanotechnol.* **2019**, *19*, 2154–2157. [CrossRef] [PubMed]
34. Shang, H.; Mohanram, A.; Olevsky, E.; Bordia, R.K. Evolution of anisotropy in hierarchical porous ceramics during sinter-forging. *J. Eur. Ceram. Soc.* **2016**, *36*, 2937–2945. [CrossRef]
35. Lame, O.; Bouvard, D.; Wiedemann, H. Anisotropic shrinkage and gravity induced creep during sintering of steel powder compacts. *Powder Metall.* **2002**, *45*, 181–185. [CrossRef]
36. Sato, K.; Yilmaz, H.; Hotta, Y.; Ijuin, A.; Watari, K. Dispersion of Ceramic Particles in Aqueous Media with Surface-Grafted Dispersant. *J. Am. Ceram. Soc.* **2009**, *92*, 256–259. [CrossRef]
37. Hu, C.; Chen, Y.; Yang, T.; Liu, H.; Huang, X.; Huo, Y.; Jia, Z.; Wang, H.; Hu, L.; Sun, H. Effect of SiC powder on the properties of SiC slurry for stereolithography. *Ceram. Int.* **2021**, *47*, 12442–12449. [CrossRef]
38. Manière, C.; Kerbart, G.; Harnois, C.; Marinel, S. Modeling sintering anisotropy in ceramic stereolithography of silica. *Acta Mater.* **2020**, *182*, 163–171. [CrossRef]

materials

Article

Optimized Zirconia 3D Printing Using Digital Light Processing with Continuous Film Supply and Recyclable Slurry System

Waqas Ahmed Sarwar [1], Jin-Ho Kang [2] and Hyung-In Yoon [1,*]

[1] Department of Prosthodontics, School of Dentistry and Dental Research Institute, Seoul National University, Seoul 03080, Korea; engr.raowaqas@gmail.com
[2] Department of Prosthodontics, School of Dentistry, Chonnam National University, Gwangju 61186, Korea; jhk.bme1002@gmail.com
* Correspondence: drhiy226@snu.ac.kr; Tel.: +82-2-2072-4472

Abstract: Stereolithography (SL) can fabricate complex ceramic parts layer by layer using computer-aided design (CAD) models. The traditional SL system utilizes a vat filled with ceramic slurry with a high solid content, which for ceramics contributes to several limitations and operational difficulties, and further renders it nonrecyclable mainly due to the presence of printed residue and its high viscosity. In this study, we utilized a continuous film supply (CFS) system integrated with a tape-casting type digital light processing (DLP) printer to fabricate zirconia prototypes with a solid content of 45 volume percent (vol.%). Various printing and postprocessing parameters were studied for optimization, to achieve a relative density of 99.02% ± 0.08% with a microhardness of 12.59 ± 0.47 GPa. Slurry reusability was also demonstrated by printing with recycled slurry to produce consistent relative density values in the range of 98.86% ± 0.02% to 98.94% ± 0.03%. This method provides new opportunities for material recycling and the fabrication of dense complex ceramic products, reducing the consumption of the material.

Keywords: zirconia; digital light processing; continuous film supply; recyclable slurry

Citation: Sarwar, W.A.; Kang, J.-H.; Yoon, H.-I. Optimized Zirconia 3D Printing Using Digital Light Processing with Continuous Film Supply and Recyclable Slurry System. *Materials* **2021**, *14*, 3446. https://doi.org/10.3390/ma14133446

Academic Editor: Matthias Kern

Received: 20 May 2021
Accepted: 18 June 2021
Published: 22 June 2021

1. Introduction

Zirconia ceramics provide unique properties in the field of biomedical applications. They demonstrate a low affinity to bacterial plaque, limit inflammatory infiltration, and provide good soft-tissue integration. However, compared to metals and polymers, high-strength ceramics such as zirconia are relatively difficult to process into intricate and complex structures via subtractive machining. Additive manufacturing (AM) fabricates three-dimensional (3D) objects based on computer-aided design (CAD) models by depositing printable materials layer by layer. The AM process can produce a variety of complex and intricate geometrical shapes simultaneously in a single workflow, without wasting machining tools and production time. With all these benefits, additively manufactured ceramic products are available using various technologies such as binder jetting, powder bed fusion, material extrusion, and vat polymerization [1–3]. Stereolithography (SL) or digital light processing (DLP) projector-based vat polymerization has especially high production accuracy and an excellent surface finish [4–8].

In vat-polymerization-based AM processes using ceramics, a high solid loading is required to fabricate fully dense ceramic objects, causing the aggregation of ceramic particles in the slurry and resulting in a high viscosity [9–11]. High viscosity produces challenges in maintaining accurate layer thickness and proper degassing. Although viscosity can be reduced by the use of diluents or temperature control, this is limited, as diluents can have adverse effects on the final density and produce high sintering shrinkage [12–17]. In addition, the recoating of the ceramic layer material during the printing recoating requires slurry of a low viscosity [18–21]. The homogenous temperature control of the whole vat area may help control the viscosity, though this is a challenging process [22].

Another major issue in vat polymerization with ceramics is light scattering by ceramic particles, resulting in the unwanted overgrowth of features that not only affects cure depth but also causes unexpected curing in bilateral directions [15,23–27]. Furthermore, the green printed state of ceramic material is fragile and lacking in the strength required to absorb the stresses involved in printing, making it susceptible to breakage during printing [28]. The presence of detached cured layers or partially failed samples in the vat poses hindrances for further printing, making the remaining slurry potentially non-reusable. Overall, vat polymerization faces many challenges in the slurry dispensing system, which cannot accommodate a higher viscosity without wasting slurry and processing time [29]. As an alternative method to vat polymerization, tape-casting-based additive manufacturing [22,24,28] can be utilized to build up ceramic materials layer by layer. It uses a casting head that acts as both a slurry reservoir and a material dispensing system. It is usually comprised of a single doctor blade that shears the slurry into a thin film onto a tape passing below the blades, providing a continuous film with uniform thickness [28]. It can also accommodate relatively higher-viscosity ceramics [28].

The purpose of this study was to produce high-density zirconia prototypes using a tape-casting type DLP printer with a continuous film supply by the optimization of processing parameters, and to evaluate the recycling efficiency of the system. The zirconia ceramic slurry used in this AM was characterized and processed for the complete printing workflow. The viscosity, film thickness, slurry recycling efficiency, thermogravimetry, shrinkage rate, density, and microhardness were assessed to evaluate the performance of the printing system.

2. Materials and Methods

2.1. Printer Design and Processing

A modified version of a Onestage 6500 (Illuminaid Inc., Seoul, Korea) printer with a DLP projector was utilized for this research. The printer provides a continuous film of slurry to be printed in layers and is referred to as the "continuous film supply (CFS) system" throughout this study. The schematic design for this printing machine is shown in Figure 1. It mainly consists of a transparent hydrophobic tape conveyor coated with silicone material that moves from one roller to the other, passing through a casting head with dual doctor blades, a build platform, and a recycling blade. The film recoating thickness is controlled by physically controlling the double doctor blades and rolling speed. The projector installed in the DLP has an LED light source with a maximum irradiance of 5.19 mW/cm^2 at a peak wavelength of 405 nm.

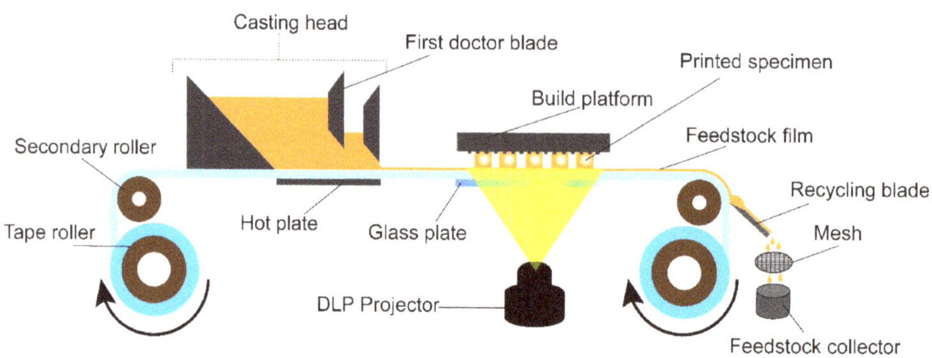

Figure 1. Schematic of the continuous film supply system with digital light processing during the three-dimensional printing process.

The system worked in sequences running successively, one after another. The first sequence initiated the rollers to move the tape for a set distance at a controlled speed, such that the slurry-recoated film arrived just under the build platform. The second sequence moved the build stage down to a set distance at a controlled speed, and decelerated close to its destination on the glass plate. The third sequence controlled the curing times for each designated set of layers of the whole process. The fourth sequence moved the platform up and the glass plate down, and finally, the manufacturing process repeated from the first sequence again. After hardening the selective area of the layer on the platform, the unused slurry was left on the tape. The unused slurry was peeled off from the tape using the recycling blade, where slurry flowed downward into the recycled-slurry collector after passing through a mesh of 100 μm to remove any agglomerations produced during printing, such as detached layers or damaged samples.

2.2. Slurry Preparation

A solvent-free photocurable ceramic suspension was formulated, containing high-purity ceramic powder dispersed in an organic resin composed of two multifunctional acrylate monomers, two photoinitiators, and some processing additives. A 3 mol % yttria-stabilized zirconia powder (TZ-3YS-E, Tosoh Corp, Tokyo, Japan) with an average particle size of 90 nm was used for this slurry. The zirconia ceramic powder was coated with a dispersant by ball milling in ethanol for 5 h and then dried in a vacuum oven at 30 °C. A nonvolatile hydroxy-functional ester-based dispersant was chosen for the zirconia ceramic suspension, based on the surface chemistry of the ceramic powder and resin. For this research, a ceramic suspension with a solid loading of 45 vol.% was prepared to balance the required density and printable viscosity. A suspension with a solid loading of 44.65 vol.% was chosen for the recycling experiment to ease the filtration and recycling process. The prepared ceramic suspension (slurry) was further homogenized before printing with a planetary centrifugal mixer (Thinky, ARE-310, Tokyo, Japan). The degassing process was then conducted in a −0.1 MPa vacuum for 5 min. The homogenization and degassing processes were also carried out for the recycled slurry.

2.3. Specimen 3D Printing

The 3D printing of each specimen was performed according to the sequence mentioned above. The layer thickness was kept at 50 μm for this research. An exposure energy of 20.76 mJ/cm^2 was applied to print each layer except for the first five layers, to which 62.28 mJ/cm^2 was applied [30,31]. Other parameters used are mentioned in Table 1. The wait time mentioned in the table refers to the time the build platform spent on the slurry film just before and after curing, without changing its position. A standard tessellation language (STL) file format was used as a CAD model (Figure 2) with specimen dimensions of 7 × 7 × 4 mm^3 scaled linearly in all dimensions, considering compensation for the overgrowth percentage and sintering shrinkage.

Table 1. Several parameters used for printing.

Parameter	Value
Wait time before curing	12 s
Wait time after curing	5 s
Tape speed	8 mm/s
Build platform speed (z-direction)	1 mm/s
Height of first blade	1 mm
Height of second blade	0.2 mm

Figure 2. A standard tessellation language file image for the standard specimens to be printed.

2.4. Postprocessing

All the printed specimens were cleaned using isopropanol through a handheld atomizing spray. Debinding was performed using an atmosphere type 1 furnace (PT-17EF022, PyroTech, Gyeonggi-do, Suwon si, Korea) in an argon atmosphere with a heating rate of 10 °C/h until 600 °C, with a dwell time of 2 h. A stairway approach to 600 °C was used with several holding temperatures, following derivative thermogravimetry analysis data. The sintering process was conducted in air at 1550 °C for 2 h with a heating rate of 2.5 °C/h, shown in Figure 3, using a Super Kanthal furnace (AJ-SKB6, Ajeon, Gyeonggi-do, Suwon si, Korea). The weight percent values of the solid loading of the specimens printed using the original ceramic slurry and the recycled slurry were measured from the weight difference between the specimens after debinding and sintering.

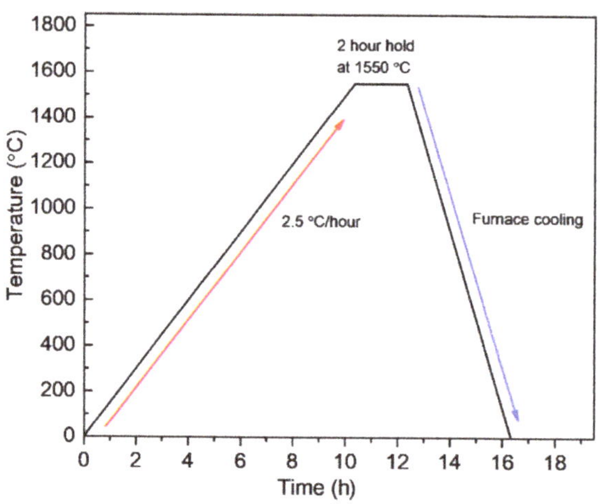

Figure 3. Sintering schedule followed by an atmosphere post debinding process in air.

2.5. Viscosity Analysis

The viscosity of 2 mL of the ceramic slurry after degassing was measured at 30, 40, 50, and 60 °C using a rotational rheometer (TA Instruments, ARES-G2, New Castle, DE, USA).

2.6. *Film Thickness Analysis of Slurry*

The film thickness was measured using a micrometer (293-240-30, Mitutoyo, Kawasaki, Japan) with an accuracy of ±1 μm. The whole film above the projector was hardened and an average value was taken after taking measurements across the width. Measurement was repeated 5 times to calculate standard deviation and maintain a consistent slurry amount behind the blade.

2.7. *Thermogravimetric Analysis*

Simultaneous thermogravimetric–derivative thermogravimetry analysis (TG-DTG) was conducted using an analyzer device (SDT-Q600, TA Instrument, New Castle, DE, USA) on a printed cubic specimen with a size of $3 \times 3 \times 3$ mm^3, in argon atmosphere, and with a 1 °C/min temperature gradient from room temperature to 600 °C.

2.8. *Density Analysis*

The density of the printed specimen was characterized via the Archimedean principle using 10 replicates and an analytical balance (Adam Equipment, SAB 125i, Oxford, CT, USA) with readability of 10 μg. For recycled slurry, five replicates were used due to the limited amount of recycled slurry.

2.9. *Microhardness Testing*

The microhardness test value was measured by determining the Vickers hardness (Shimadzu, HMV-2, Kyoto, Japan) on mirror-polished cross-sections of the specimens, according to ASTM C1327-15 [32]. A loading force of 9.81 N was applied for 10 s, and the diagonals were measured through a ×40 objective lens. A total of 10 specimens were tested with at least 5 indents each, and the average value for those indents was considered as the final hardness value of each specimen.

2.10. *Microstructure Analysis*

The microstructure analysis was performed through a field-emission scanning electron microscope (AURIGA, Carl Zeiss, Oberkochen, Germany) on platinum-coated specimens with a coating thickness of 5 nm. The analysis was carried out on the fine-polished cross-sections of both as-printed specimens and sintered specimens after thermal etching at 1450 °C for 40 min. The grain size was measured for 5 sintered samples, printed in separate batches, by the linear intercept method with a conversion factor of 1.56 [33].

3. Results and Discussion

3.1. *Viscosity Analysis*

Slurry with a 45 vol.% solid loading of zirconia coated with dispersant was tested at four different temperatures; however, it was found that the slurry started to thermally cure at temperatures higher than 60 °C. At around 60 °C, the slurry started to show signs of thermal curing and formed agglomerates which started to collect in the recycling mesh. The viscosity seemed to decrease exponentially with rising temperatures (Figure 4a). Viscosity decreased by a factor of four from 57 Pa·s at 30 °C to 13 Pa·s at 50 °C. The effect of the dispersant also played a role in this decrease in viscosity at higher temperatures, as the dispersant is most effective at temperatures higher than 30 °C. A working temperature of 50 °C was chosen, considering the lowest available viscosity without the possibility of the thermal curing of the slurry. The doctor blade requires a suitable viscosity range to make a homogenous film. To control and increase the printable viscosity range, temperature control and a proper dispersant of an optimized amount are essential [22,34]. The casting head, which acts as a slurry reservoir, has a hot plate below it to control temperature. Furthermore, a diluent can also be used to drastically reduce the viscosity; however, to avoid higher sintering shrinkage and to retain sufficient viscosity for the doctor blade, a diluent was not used.

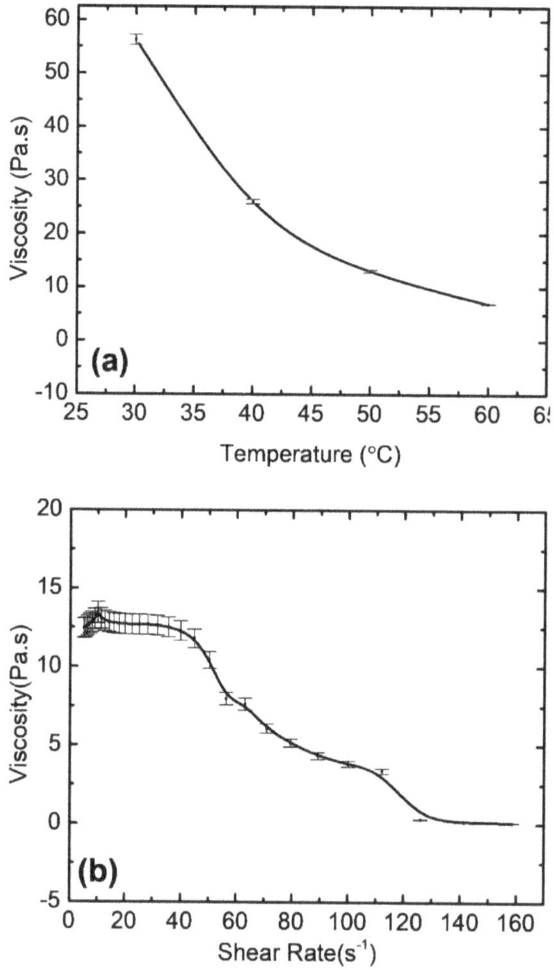

Figure 4. (**a**) Change in viscosity with the temperature at a 10 s^{-1} shear rate; (**b**) effect of shear rate on the viscosity of zirconia slurry at 50 °C.

At 50 °C, the viscosity was also tested with varying shear rates (Figure 4b). The viscosity decreased from 12.26 × 103 mPa·s at 3 s^{-1} to 96.95 mPa·s at 160 s^{-1}. The flow curve was consistent with non-Newtonian flow behavior and showed pseudoplastic behavior. Considering that the tape velocity under the casting head was 8 mm/s, the viscosity was found to be in a suitable range for the doctor blades to sufficiently coat the slurry film at a uniform thickness.

3.2. *Film Thickness of Slurry*

Figure 5 shows the effect of the amount of slurry in the casting head behind the blades on the coated film thickness. It shows two trends, with the first trend (dotted) starting with 100% of the slurry reservoir filled with slurry and the second trend (solid) starting from 10% slurry. The dotted line depicts the natural process of recoating during printing with the slurry amount reducing as printing continues. The solid line depicts a trend when, at any point during printing, the reservoir was filled again with fresh or recycled slurry. Considering the dotted line, the film thickness stayed consistent with a mean difference of 5 ± 1 μm from slurry amounts of 100% (149 ± 2 μm) to 25% (154 ± 2 μm). For the solid line, the mean film thickness was between 86 ± 3 μm and 90 ± 3 μm, when the slurry amount varied from 20% to 100%, respectively. If a partial area was observed in the case of refill, the mean difference in film thickness was 4 ± 2 μm.

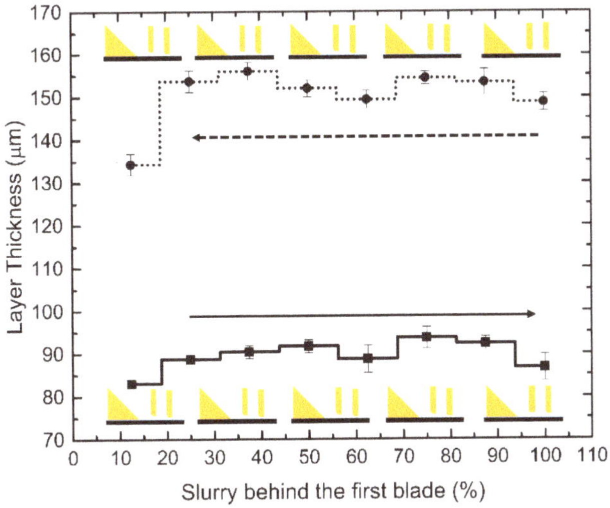

Figure 5. Film thickness measurement on the tape versus slurry level behind the first doctor blade. First with ascending level (**solid line**); second with descending level (**dotted line**).

Film thickness depends on the viscosity, hydrodynamic pressure, tape speed, width of the blades, and density of the slurry [24,35,36]. Viscosity plays an important role and can change the flow pattern of slurry in diverse ways; however, with a consistent viscosity, the flow pattern changes with both Newtonian and non-Newtonian fluids only slightly [37]. In this study, the CFS utilized a dual-blade casting head, which in cases of variation in the slurry amount behind the first blade causes a minor change in film thickness if a sufficient tape velocity is maintained [38,39]. However, in the case of a non-steady state, the effect of slurry addition for the completion of the printing process (recycling) on the film thickness is also important. This addition of slurry, at any time, causes a sudden change in hydrodynamic pressure behind the blades, which can eventually affect film thickness. The effect of refilling the casting head was minute changes in film thickness unless it fell below a certain slurry height, which, in our case, was determined to be less than 25% (10 mm) of the total capacity. Layer thickness was controlled by the position of the build platform before curing. The build platform maintained its height for the programmed layer thickness. Recoated film was always fabricated 50–80 μm higher than the programmed layer thickness for compensation, which was then pushed out by compression by the build plate, to cure the exact layer thickness. Considering this alternative film recoating system of the doctor blades, the accurate layer thickness could be maintained in highly viscous slurry conditions, even when the slurry was below 10% capacity of the casting head, by the use of compensation. The excess slurry lost on the film was recovered by recycling.

3.3. *Thermogravimetric Analysis*

Postprocessing after cleaning and drying started with binder burnout, which was carried out separately from sintering to check the effectiveness of the binder burnout and diagnose the initiation of possible thermal cracks. The derivative of weight (DW) curve shown in Figure 6 exhibits a sequence of peaks that indicate the presence of many phase changes during thermal degradation. Even though there are three distinctive peaks between 300 and 500 °C, their lower portions overlap with each other, which could, in some cases, hide smaller peaks in the overlapped regions. This makes the debinding process in this temperature zone heat sensitive, and it requires a comparatively lower heating rate. Considering the overlapping peaks in the DW curve, a heating rate of 10 °C/h in argon was utilized to eliminate thermal cracks, as an air atmosphere was found to cause thermal cracks after debinding, even at a slower heating rate of 5 °C/h.

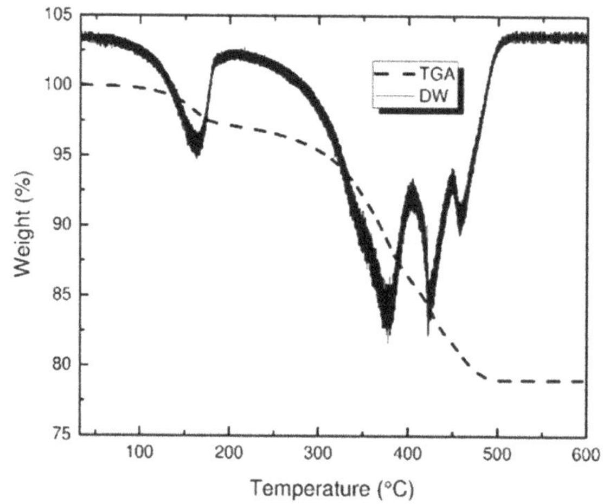

Figure 6. Thermogravimetric analysis representing both weight change percent and the rate of change in weight vs. temperature in an argon atmosphere.

Oxidative degradation produced mainly gaseous products that, under hindered diffusion, can create microcracks and blisters due to the pressure buildup of gas inside the sample. In the case of thermal degradation, polymer chains break down by chain scission with byproducts including carbon black, carbon monoxide, ammonia aliphatic amines, ketones, nitriles, and hydrogen cyanide [40]. Thermal degradation involves the removal of functional groups one after the other to leave behind only carbon black. Thermal degradation is a much slower process, and can be considered more sensitive as compared to oxidative degradation when crack formation during post-processes is compared. To avoid thermal stresses and cracks in sintered products, debinding was performed in an argon environment.

3.4. Densification

After cleaning, drying was performed at room temperature and no observable drying shrinkage was detected. In this study, 45 vol.% was the optimum solid loading in the argon atmosphere to achieve a relative densification of 99.02% ± 0.08% (5.99 ± 0.01 g/cm^3) for zirconia. Linear shrinkage is shown in Table 2, which indicates homogenous sintering with shape retention and a lack of deformation during post-processing. Uniform sintering shrinkage in all dimensions is essential for any application, especially to produce an accurate final product. The shrinkage rate of 20–25% was within the normal limits for traditional CAD/CAM pre-sintered zirconia blocks [41,42].

Table 2. Sintering shrinkage in all directions.

X-Direction (%)	Y-Direction (%)	Z-Direction (%)
22.59 ± 0.19	22.48 ± 0.15	22.56 ± 0.11

3.5. Slurry Recycling Efficiency

All the residual slurry remaining on the tape was recycled after filtration and could be recycled again and used to refill the casting head after degassing for continuous printing.

The characterization of samples printed using recycled slurry is given in Table 3. The results showed that there was a change in solid loading of 0.12 weight percent for the first recycling instance, and 0.26 weight percent for the second recycling instance. Changes in solid loading for recycled slurry can be explained by considering scattering agglomeration, which is the curing of slurry present near the exposed light area by light scattered by ceramic powder [24,43,44]. However, such a minute change in weight percent in a high-solid-load slurry was negligible. This was evident in the relative density of samples prepared only by using recycled slurry unmixed with fresh slurry, as there was only a slight change in density, making printing possible even after two instances of recycling. In practice, recycled slurry is mixed with the fresh slurry already present in the casting head, which can reduce solid loading changes further.

Table 3. Change in properties with recycling.

Properties	First Use	Recycled Once	Recycled Twice	Recycled Thrice
Weight percent (%)	81.32 ± 0.03	81.44 ± 0.04	81.58 ± 0.06	81.61 ± 0.02
Relative Density (%)	98.89 ± 0.03	98.86 ± 0.02	98.94 ± 0.03	98.91 ± 0.03

3.6. Microhardness Test

Sintered specimens showed an average microhardness of 12.59 ± 0.47 GPa. As observed from Figure 7, the shape of the indent was symmetrical, the cracks produced at the corners of the square pyramid were linear, and it was well within the range of acceptable indentation by ASTM. Only indents producing Palmqvist cracks were measured for diagonals, and unsymmetrical indents were ignored during the measurement. The indentation seemed to include microporosity, as seen inside the pyramid; however, the porosity was not large enough to deform the indentation. A microhardness of over 12 GPa at 45 vol.% solid loading was well within the reported value produced by the additive manufacturing and conventional subtractive machining of zirconia plates [45,46].

Figure 7. Indent at 1000× magnification, for reference.

3.7. Microstructure Analysis

The sintered cross-section did not show any signs of boundaries between layers, which were visible before debinding, as seen in Figure 8. This disappearance of layer boundaries after sintering is not common among traditional vat-polymerization printed parts. The disappearance of layer boundaries, after sintering and attaining a relative density of more than 99% at 45 vol.%, showed that tape casting produced compact as-printed objects. The DLP with CFS system showed promise in producing dense structures. The as-printed specimen is shown in Figure 9a as an example of printability. Furthermore, at 30,000× magnification the microstructure showed a mean grain size of 644 ± 20 nm and was densely packed, as evident in Figure 9b. However, further refinement of the grain size could be achieved by increasing solid loading and decreasing the sintering temperature.

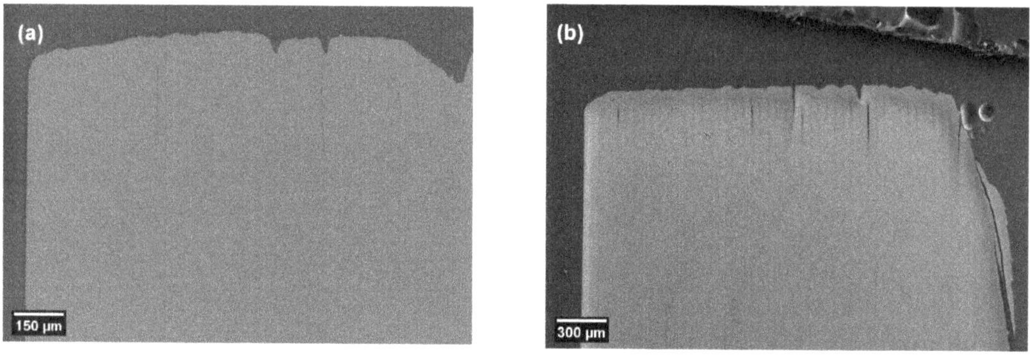

Figure 8. (**a**) Cross-section of sintered specimen at 200× (back-scattered electron mode) to observe any layer boundaries; (**b**) cross-section of as-printed specimen at 100× to observe distinctive layer boundaries.

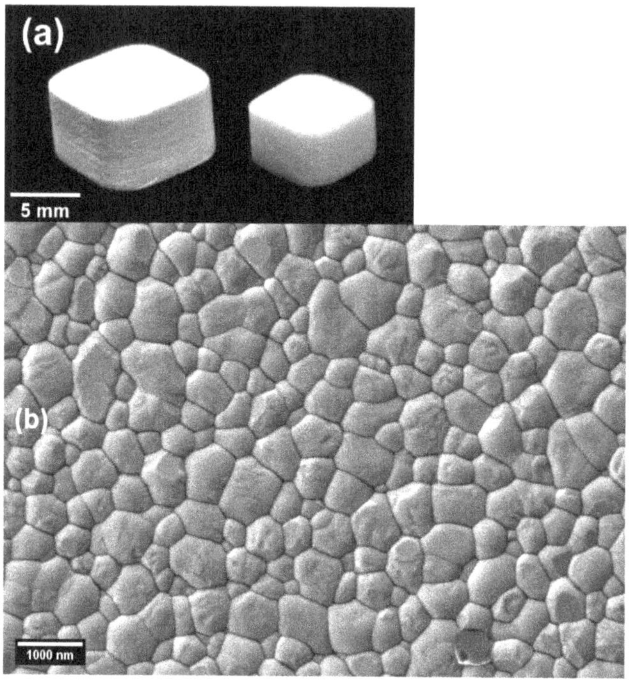

Figure 9. (**a**) As-printed and sintered samples for reference; (**b**) microstructure of the polished cross-section of 45 vol.% solid loaded zirconia at 30,000×.

4. Conclusions

Based on this study, the tape casting DLP printing with CFS is a capable alternative to vat polymerization to print dense zirconia prototypes using high-viscosity slurry. An appropriate dispersant and working temperature were necessary to achieve printable viscosity. Consistency of film thickness on the tape, throughout the whole printing process, was imperative for the DLP with CFS. Dual doctor blades were found to produce a film thickness well above the overall set layer thickness. Using the DLP with CFS, a 45 vol.%

solid loaded 3Y-TZP slurry was used to fabricate ceramic products of a homogenous microstructure, with a relative density of 99.02% ± 0.08% and a mean grain size of 644 ± 20 nm after postprocessing. The microhardness value of the ceramic product was 12.59 ± 0.47 GPa. Furthermore, the DLP with CFS can effectively recycle and reuse slurry, providing a consistent density for at least two instances of recycling, making it a competitive alternative to conventional processing techniques for the fabrication of dense ceramic products.

Author Contributions: Conceptualization, H.-I.Y.; methodology, H.-I.Y., W.A.S. and J.-H.K.; formal analysis, W.A.S.; writing—original draft preparation, W.A.S.; writing—review and editing, H.-I.Y., W.A.S. and J.-H.K.; supervision, H.-I.Y.; funding acquisition, H.-I.Y. All authors have read and agreed to the published version of the manuscript.

Funding: This work was supported by Creative-Pioneering Researchers Program through Seoul National University (SNU, 860-20200093).

Institutional Review Board Statement: Not applicable.

Informed Consent Statement: Not applicable.

Data Availability Statement: The data presented in this study are available on request from the corresponding author.

Conflicts of Interest: The authors declare no conflict of interest.

References

1. Jang, K.J.; Kang, J.H.; Fisher, J.G.; Park, S.W. Effect of the volume fraction of zirconia suspensions on the microstructure and physical properties of products produced by additive manufacturing. *Dent. Mater.* **2019**, *35*, e97–e106. [CrossRef] [PubMed]
2. Kang, J.H.; Kaneda, J.; Jang, J.G.; Sakthiabirami, K.; Lui, E.; Kim, C.; Wang, A.; Park, S.W.; Yang, Y.P. The Influence of Electron Beam Sterilization on In Vivo Degradation of beta-TCP/PCL of Different Composite Ratios for Bone Tissue Engineering. *Micromachines* **2020**, *11*, 273. [CrossRef] [PubMed]
3. Kang, J.-H.; Jang, K.-J.; Sakthiabirami, K.; Oh, G.-J.; Jang, J.-G.; Park, C.; Lim, H.-P.; Yun, K.-D.; Park, S.-W. Mechanical properties and optical evaluation of scaffolds produced from 45S5 bioactive glass suspensions via stereolithography. *Ceram. Int.* **2020**, *46*, 2481–2488. [CrossRef]
4. Ruiz-Morales, J.C.; Tarancón, A.; Canales-Vázquez, J.; Méndez-Ramos, J.; Hernández-Afonso, L.; Acosta-Mora, P.; Marín Rueda, J.R.; Fernández-González, R. Three dimensional printing of components and functional devices for energy and environmental applications. *Energy Environ. Sci.* **2017**, *10*, 846–859. [CrossRef]
5. Hernández, E.; Acosta-Mora, P.; Méndez-Ramos, J.; Borges, M.; Esparza, P.; Canales-Vázquez, J.; Núñez, P.; Ruiz-Morales, J.C. Prospective use of the 3D printing technology for the microstructural engineering of Solid Oxide Fuel Cell components. *Comput. Aided Des.* **2014**, *53*, 213–216. [CrossRef]
6. Tamayo-Domínguez, A.; González, J.M.F.; Sierra-Castañer, M. Stereolithography and Direct Metal Laser Sintering Applied to mm-Wave Antennas. In Proceedings of the 2020 International Symposium on Antennas and Propagation (ISAP), Virtual Conference, 25–28 January 2021; pp. 273–274.
7. Wang, W.; Sun, J. Dimensional accuracy and clinical adaptation of ceramic crowns fabricated with the stereolithography technique. *J. Prosthet. Dent.* **2021**, *125*, 657–663. [CrossRef] [PubMed]
8. Xing, H.; Zou, B.; Liu, X.; Wang, X.; Huang, C.; Hu, Y. Fabrication strategy of complicated Al2O3-Si3N4 functionally graded materials by stereolithography 3D printing. *J. Eur. Ceram. Soc.* **2020**, *40*, 5797–5809. [CrossRef]
9. Zhang, K.; Meng, Q.; Cai, N.; Qu, Z.; He, R. Effects of solid loading on stereolithographic additive manufactured ZrO2 ceramic: A quantitative defect study by X-ray computed tomography. *Ceram. Int.* **2021**. [CrossRef]
10. Hu, K.; Wei, Y.; Lu, Z.; Wan, L.; Li, P. Design of a Shaping System for Stereolithography with High Solid Loading Ceramic Suspensions. *3D Print. Addit. Manuf.* **2018**, *5*, 311–318. [CrossRef]
11. Zhang, K.; Xie, C.; Wang, G.; He, R.; Ding, G.; Wang, M.; Dai, D.; Fang, D. High solid loading, low viscosity photosensitive Al_2O_3 slurry for stereolithography based additive manufacturing. *Ceram. Int.* **2019**, *45*, 203–208. [CrossRef]
12. Li, K.; Zhao, Z. The effect of the surfactants on the formulation of UV-curable SLA alumina suspension. *Ceram. Int.* **2017**, *43*, 4761–4767. [CrossRef]
13. Wang, Z.; Huang, C.; Wang, J.; Zou, B. Development of a novel aqueous hydroxyapatite suspension for stereolithography applied to bone tissue engineering. *Ceram. Int.* **2019**, *45*, 3902–3909. [CrossRef]
14. Lee, Y.-H.; Lee, J.-B.; Maeng, W.-Y.; Koh, Y.-H.; Kim, H.-E. Photocurable ceramic slurry using solid camphor as novel diluent for conventional digital light processing (DLP) process. *J. Eur. Ceram. Soc.* **2019**, *39*, 4358–4365. [CrossRef]
15. Johansson, E.; Lidström, O.; Johansson, J.; Lyckfeldt, O.; Adolfsson, E. Influence of Resin Composition on the Defect Formation in Alumina Manufactured by Stereolithography. *Materials* **2017**, *10*, 138. [CrossRef] [PubMed]

16. Melchels, F.P.W.; Feijen, J.; Grijpma, D.W. A poly(d,l-lactide) resin for the preparation of tissue engineering scaffolds by stereolithography. *Biomaterials* **2009**, *30*, 3801–3809. [CrossRef]
17. Jang, J.H.; Wang, S.; Pilgrim, S.M.; Schulze, W.A. Preparation and characterization of barium titanate suspensions for stereolithography. *J. Am. Ceram. Soc.* **2000**, *83*, 1804–1806. [CrossRef]
18. Goswami, A.; Ankit, K.; Balashanmugam, N.; Umarji, A.M.; Madras, G. Optimization of rheological properties of photopolymerizable alumina suspensions for ceramic microstereolithography. *Ceram. Int.* **2014**, *40*, 3655–3665. [CrossRef]
19. Janusziewicz, R.; Tumbleston, J.R.; Quintanilla, A.L.; Mecham, S.J.; DeSimone, J.M. Layerless fabrication with continuous liquid interface production. *Proc. Natl. Acad. Sci. USA* **2016**, *113*, 11703–11708. [CrossRef] [PubMed]
20. Li, X.; Wang, M.; Xing, L.; Chu, J.; Bao, Y. *Experimental and Numerical Investigation on Surface Damage of Cold Rolled Sheet Caused by Inclusion Movement*; Springer International Publishing: Cham, Switzerland, 2020; pp. 239–247.
21. Wu, Z.; Liu, W.; Wu, H.; Huang, R.; He, R.; Jiang, Q.; Chen, Y.; Ji, X.; Tian, Z.; Wu, S. Research into the mechanical properties, sintering mechanism and microstructure evolution of $Al2O_3$-ZrO_2 composites fabricated by a stereolithography-based 3D printing method. *Mater. Chem. Phys.* **2018**, *207*, 1–10. [CrossRef]
22. Hinczewski, C.; Corbel, S.; Chartier, T. Ceramic suspensions suitable for stereolithography. *J. Eur. Ceram. Soc.* **1998**, *18*, 583–590. [CrossRef]
23. O'Masta, M.R.; Stonkevitch, E.; Porter, K.A.; Bui, P.P.; Eckel, Z.C.; Schaedler, T.A. Additive manufacturing of polymer-derived ceramic matrix composites. *J. Am. Ceram. Soc.* **2020**, *103*, 6712–6723. [CrossRef]
24. Alazzawi, M.K.; Beyoglu, B.; Haber, R.A. A study in a tape casting based stereolithography apparatus: Role of layer thickness and casting shear rate. *J. Manuf. Process.* **2021**, *64*, 1196–1203. [CrossRef]
25. Halloran, J.W. Ceramic Stereolithography: Additive Manufacturing for Ceramics by Photopolymerization. *Annu. Rev. Mater. Res.* **2016**, *46*, 19–40. [CrossRef]
26. Gentry, S.P.; Halloran, J.W. Depth and width of cured lines in photopolymerizable ceramic suspensions. *J. Eur. Ceram. Soc.* **2013**, *33*, 1981–1988. [CrossRef]
27. Gentry, S.P.; Halloran, J.W. Light scattering in absorbing ceramic suspensions: Effect on the width and depth of photopolymerized features. *J. Eur. Ceram. Soc.* **2015**, *35*, 1895–1904. [CrossRef]
28. Song, X.; Chen, Y.; Lee, T.W.; Wu, S.; Cheng, L. Ceramic fabrication using Mask-Image-Projection-based Stereolithography integrated with tape-casting. *J. Manuf. Process.* **2015**, *20*, 456–464. [CrossRef]
29. Song, X. *Slurry Based Stereolithography: A Solid Freeform Fabrication Method of Ceramics and Composites*; University of Southern California: Los Angeles, CA, USA, 2016.
30. Mills, P.; Raymont, J. Ultraviolet (UV) measurement for formulators: Park I. *RadTech Rep.* **2009**, *23*, 34.
31. Conti, L.; Bienenstein, D.; Borlaf, M.; Graule, T. Effects of the Layer Height and Exposure Energy on the Lateral Resolution of Zirconia Parts Printed by Lithography-Based Additive Manufacturing. *Materials* **2020**, *13*, 1317. [CrossRef]
32. ASTM C1327-15(2019). *Standard Test Method for Vickers Indentation Hardness of Advanced Ceramics*; ASTM International: West Conshohocken, PA, USA, 2019. [CrossRef]
33. Mendelson, M.I. Average Grain Size in Polycrystalline Ceramics. *J. Am. Ceram. Soc.* **1969**, *52*, 443–446. [CrossRef]
34. Robert, G.O.; Timothy, N.P. *Computational Rheology*; Imperial College Press: London, UK, 2002.
35. Chou, Y.T.; Ko, Y.T.; Yan, M.F. Fluid flow model for ceramic tape casting. *J. Am. Ceram. Soc.* **1987**, *70*, C-280–C-282. [CrossRef]
36. Huang, X.; Liu, C.; Gong, H. A viscoplastic flow modeling of ceramic tape casting. *Mater. Manuf. Process.* **1997**, *12*, 935–943. [CrossRef]
37. Wonisch, A.; Polfer, P.; Kraft, T.; Dellert, A.; Heunisch, A.; Roosen, A. A Comprehensive Simulation Scheme for Tape Casting: From Flow Behavior to Anisotropy Development. *J. Am. Ceram. Soc.* **2011**, *94*, 2053–2060. [CrossRef]
38. Zhang, G.; Wang, Y.; Ma, J. Bingham plastic fluid flow model for ceramic tape casting. *Mater. Sci. Eng. A* **2002**, *337*, 274–280. [CrossRef]
39. Jabbari, M.; Hattel, J. Bingham plastic fluid flow model in tape casting of ceramics using two doctor blades—Analytical approach. *Mater. Sci. Technol.* **2014**, *30*, 283–288. [CrossRef]
40. Trunec, M.; Cihlar, J. Removal of thermoplastic binders from ceramic green bodies. *Ceramics* **1997**, *41*, 67–80.
41. Raigrodski, A.J. Contemporary all-ceramic fixed partial dentures: A review. *Dent. Clin. N. Am.* **2004**, *48*, 531–544. [CrossRef]
42. Besimo, C.E.; Spielmann, H.P.; Rohner, H.P. Computer-assisted generation of all-ceramic crowns and fixed partial dentures. *Int. J. Comput. Dent.* **2001**, *4*, 243–262.
43. Sun, C.; Zhang, X. The influences of the material properties on ceramic micro-stereolithography. *Sens. Actuators A Phys.* **2002**, *101*, 364–370. [CrossRef]
44. Mitteramskogler, G.; Gmeiner, R.; Felzmann, R.; Gruber, S.; Hofstetter, C.; Stampfl, J.; Ebert, J.; Wachter, W.; Laubersheimer, J. Light curing strategies for lithography-based additive manufacturing of customized ceramics. *Addit. Manuf.* **2014**, *1*, 110–118. [CrossRef]
45. Cottom, B.A.; Mayo, M.J. Fracture toughness of nanocrystalline ZrO_2-3 mol% Y_2O_3 determined by Vickers indentation. *Scr. Mater.* **1996**, *34*, 809–814. [CrossRef]
46. Ruiz, L.; Readey, M.J. Effect of Heat Treatment on Grain Size, Phase Assemblage, and Mechanical Properties of 3 mol% Y-TZP. *J. Am. Ceram. Soc.* **1996**, *79*, 2331–2340. [CrossRef]

materials

MDPI

Review

Scientific Trends in Clinical Research on Zirconia Dental Implants: A Bibliometric Review

Felice Lorusso [1,2], Sammy Noumbissi [1,2], Inchingolo Francesco [3], Biagio Rapone [4], Ahmad G. A. Khater [5] and Antonio Scarano [1,2,*]

1 Department of Medical, Oral and Biotechnological Sciences, University of Chieti-Pescara, Via dei Vestini, 31, 66100 Chieti, Italy; drlorussofelice@gmail.com (F.L.); sammy@iaoci.com (S.N.)

2 Zirconia Implant Research Group (Z.I.R.G), International Academy of Ceramic Implantology, Silver Spring, MD 20910, USA

3 Department of Interdisciplinary Medicine, University of Bari Aldo Moro, 70121 Bari, Italy; f.inchingolo@icloud.com

4 Department of Basic Medical Sciences, Neurosciences and Sense Organs, University of Bari Aldo Moro, 70121 Bari, Italy; biagiorapone79@gmail.com

5 Faculty of Oral and Dental Medicine, Ahram Canadian University, 6th of October City, 8655 Giza, Egypt; ahmed.g.a.khater@gmail.com

* Correspondence: ascarano@unich.it; Tel.: +39-0871-355-4084; Fax: +39-0871-355-4099

Received: 18 October 2020; Accepted: 2 December 2020; Published: 4 December 2020

Abstract: Background: The clinical use of zirconia implants has been shown to increase steadily due to their biological, aesthetic, and physical properties; therefore, this bibliometric study aimed to review the clinical research and co-authors in the field of zirconia dental implant rehabilitation. Methods: We searched Scopus and Web of Science databases using a comprehensive search strategy to 5 October 2020, and independently paired reviewers who screened studies, and collected data with inclusion criteria restricted to clinical research only (either prospective or retrospective). Data on article title, co-authors, number of citations received, journal details, publication year, country and institution involved, funding, study design, marginal bone loss, survival rate, failure, follow-up, and the author's bibliometric data were collected and evaluated. Results: A total of 29 clinical studies were published between 2008 and 2020 as 41.4% were prospective cohort studies and 48.3% originated from Germany. Most of the included studies had been published in Clinical Oral Implant Research (n = 12), and the most productive institution was the Medical Center of University of Freiburg. The author with the largest number of clinical studies on zirconia implants was Kohal R.J. (n = 10), followed by Spies B.C. (n = 8). Conclusions: This study revealed that zirconia implants have been more prominent in the last ten years, which is a valuable option for oral rehabilitation with marginal bone loss and survival rate comparable to titanium dental implants.

Keywords: zirconia implant; bibliometrics; citations; scientometric

1. Introduction

The clinical application of dental implant rehabilitation represents consolidated effectiveness in the literature due to long-term predictability and high-level satisfactory functioning and aesthetics [1–4]. Titanium alloys are the most widely used biomaterials for dental implant fixtures due to their physical, chemical, and thermal properties, which produce the osseointegrating ability of the fixture placed to replace the natural teeth [2,5–12].

Recently, the use of zirconia as an implant material has become more prevalent due to its high aesthetic characteristics, particularly in the rehabilitation of the compromised anterior jaw area, where there is fine soft-tissue biotype and the metal sensibility of the patients [13–15].

In fact, the literature reports that the titanium ion dissolution related to the implant corrosion could alter the natural oral microbiome and the homeostatic functional balance of the oral tissues [16–20].

On the contrary, it has been shown in vitro that the zirconia surface can lead to a significant decrease in periodontal pathogen adhesion compared to the titanium surface [21], alongside similar bone–implant contact compared to the titanium fixture with an almost overlapping range [22].

Additionally, Scarano et al. reported in a rabbit study that zirconia implants had about 68.4% bone–implant contact with evidence of contact osteogenesis without fibrous tissue interposition [23].

Zirconia material is distinguished by its clear ivory appearance, which is very similar to the natural color of the teeth and is characterized by an intrinsic strength and physical resistance to the loading [24–26]; as a result, it has been introduced as a restorative material for dental crowns, bars, abutments, and specially designed drills and burs [26–34]. Therefore, zirconia has recently gained further attention in the scientific community by growing research activities to confirm the clinical effectiveness of zirconia as a dental implant material.

Although citations are not an infallible metric to determine whether research is beneficial to researchers and clinicians, citations and citation analysis can quantify an article's influence, author, subject of debate, country, journal, or a specialty [35,36]. Based on citation analysis, the bibliometric analysis aims to provide information about the trend in a research field and demonstrates its growth and development [37]; the number of citations received, researcher H-index, and journal impact factor are the most common bibliometric evaluation variables and considered as a scientific productivity score for the scientometric evaluation [38].

With the significant increase in the published articles on dental implants, recognizing trends and advances in a research field is critical and relevant to the needs of dental practitioners and researchers [39,40]. In this sense, bibliometric analysis is a useful tool for this purpose [41,42].

As far as we know, the trends and advances in zirconia dental implants have not been studied before; hence this study aimed to evaluate the bibliometric output of clinical research and co-authors in the field of zirconia dental implant rehabilitations.

2. Materials and Methods

We reported this bibliometric study in compliance with the Standards for Reporting Qualitative Research (SRQR) [43] and the Preferred Reporting Items for Systematic Reviews and Meta-Analyses (PRISMA) guidelines [44].

2.1. Search Strategy

An online literature search was conducted in Elsevier's Scopus and Clarivate Analytics' Web of Science (WoS) until 5 October 2020. We used the medical terms (MeSH) feature in the Cochrane Library to obtain the available synonyms for our search terms to create a detailed search strategy (Table 1).

Table 1. Search strategy used for each database.

Scopus	TITLE-ABS-KEY ("Zirconia" OR "Zirconium" OR "Zircon*") AND TITLE-ABS-KEY ("Dental implant" OR "Dental implants" OR "Oral implant" OR "Oral implants" OR "Implant dentistry" OR "Dental implantology" OR "Dental Implantation" OR "Osseointegrated" OR "Osseointegrated Dental Implantation") AND TITLE-ABS-KEY ("Intervention Study" OR "Clinical Trial" OR "Controlled Clinical Trial" OR "Randomized Controlled Trials OR "Non-Randomized Clinical Trial" OR "Nonrandomized Clinical Trial" OR "Quasi-Experimental" OR "Observational Study" OR "Prospective Study" OR "Prospective" OR "Retrospective Study" OR "Retrospective" OR "Comparative Study" OR "Multicenter Studies" OR "Epidemiologic Study" OR "Epidemiological Studies" OR "Cohort Study" OR "Case Studies" OR "Follow-Up Study" OR "Case-Control Study" OR "Case Report" OR "Case Series" OR "Pilot Study")

Table 1. *Cont.*

Web of Science	TS = ("Zirconia" OR "Zirconium" OR "Zircon*") AND TS = ("Dental implant" OR "Dental implants" OR "Oral implant" OR "Oral implants" OR "Implant dentistry" OR "Dental implantology" OR "Dental Implantation" OR "Osseointegrated" OR "Osseointegrated Dental Implantation") AND TS = ("Intervention Study" OR "Clinical Trial" OR "Controlled Clinical Trial OR "Randomized Controlled Trials" OR "Non-Randomized Clinical Trial" OR "Nonrandomized Clinical Trial" OR "Quasi-Experimental" OR "Observational Study" OR "Prospective Study" OR "Prospective" OR "Retrospective Study" OR "Retrospective" OR "Comparative Study" OR "Multicenter Studies" OR "Epidemiologic Study" OR "Epidemiological Studies" OR "Cohort Study" OR "Case Studies" OR "Follow-Up Study" OR "Case-Control Study" OR "Case Report" OR "Case Series" OR "Pilot Study") **Timespan:** All years. **Databases:** WOS, ARCI, BCI, KJD, MEDLINE, RSCI, SCIELO, ZOOREC. Search language = Auto.

2.2. Data Extraction and Bibliometric Parameters

We used a specially built Excel file (Microsoft, Redmond, WA, USA) to collect the findings of the literature search. The file contained the following information: abstracts, year of publication, indexed keywords, journal name, citations as well as all co-author bibliometric data (H-index, number of papers related to zirconia implant, the total number of papers, citation of paper regarding zirconia implant, and citation of paper regarding zirconia implant). Authors with the highest quantity of clinical studies regarding zirconia dental implants were evaluated and measured the average, the standard deviation, minimum and maximum of topic paper, total papers, topic citations, overall citations, and H-index. Moreover, we evaluated the scientific trend of the included study according to the year of publication and journal details (full title, the impact factor (IF), and rank) based on the Clarivate Analytics report for 2019 with selected categories: "Dentistry, Oral Surgery & Medicine", study design, number of citations received, marginal bone loss, survival rate, failure, and study follow-up.

2.3. Study Selection

We screened the literature search results in two steps, where the first phase was the screening of the title and abstract by paired reviewers separately. Then, the second phase was a full-text assessment by two expert reviewers (L.F and A.S). The reference list of the studies included in the full-text screening was hand-screened for potential additional studies. In this bibliometric study, inclusion criteria were only clinical studies (either prospective or retrospective) without time restrictions. Exclusion criteria were animal studies, in vitro studies, literature reviews, systematic reviews, short communications, personal opinion, letters, book chapters, and non-English studies.

2.4. Data Analysis

We used VOSviewer software (version 1.6.8; Leiden University, Leiden, The Netherlands) to visualize a term map analyzing keywords from the data obtained. "Create Map" function was used to analyze the data by using the "Citation" type and setting the unit of analysis as a "number of citations." In the keyword map, the node's size reflects the number of received citations, as the larger size indicates the author with the highest citations. Furthermore, keywords that often appeared together were classified as the same color in network visualization mode [45,46].

3. Results

3.1. Study Selection

A total of 1159 references were collected from electronic databases in which (n = 185) were omitted due to duplication. By title and abstract, 968 articles were screened and 841 excluded as irrelevant topics. By the full-text screening of 127 papers, 29 studies were included in this bibliometric study [47–75] excluding the remaining 98 articles because they did not meet our inclusion criteria (Figure 1).

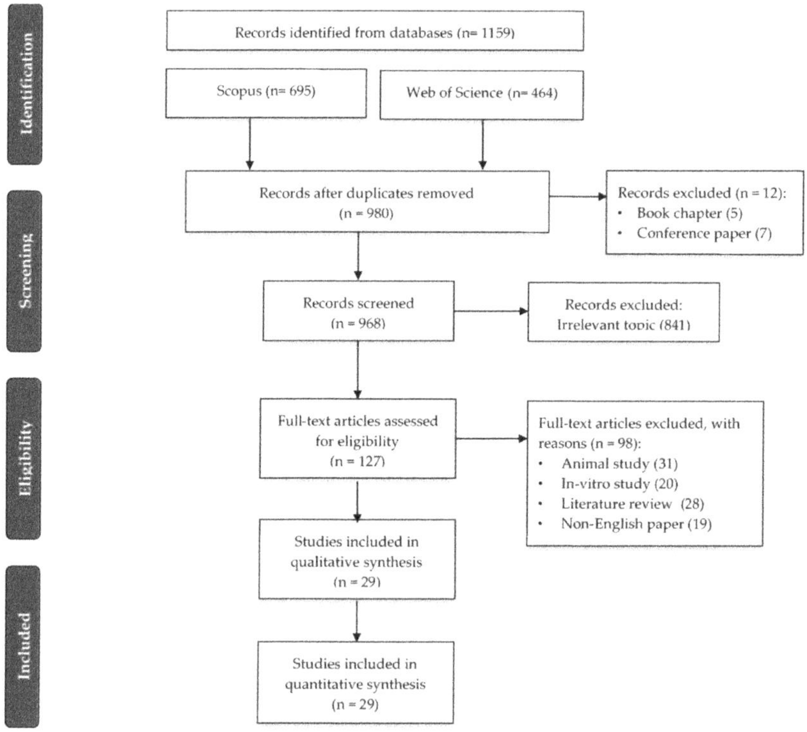

Figure 1. Preferred Reporting Items for Systematic Reviews and Meta-Analyses (PRISMA) flow chart demonstrates the process of literature search and study selection.

3.2. Study Characteristics

The included studies showed wide variability in the study design, presence/absence of a control group, experimental site, type of prosthetic rehabilitation, prosthetic connection (one-piece or two-piece), follow-up period, and different methods for evaluating the effectiveness of research. Although these differences exist, most studies reported favorable outcomes for the use of zirconia implants in oral rehabilitation. The main characteristics of the included studies are summarized in Table 2.

A total of 21 studies evaluated monolithic or one-piece zirconia implants [47–49,51–54,56–58,60, 62,64,66,70–74,76], two of which had titanium implants as their control and showed no significant difference in survival rate and marginal bone loss between groups ($p > 0.05$) [64,70]. Two studies evaluated the immediate loading of zirconia implants [53,64]: one study compared it to the non-occlusal loading procedure [64], while the other study compared it with the standard loading protocol [57]. Furthermore, 26 papers assessed the cylindrical microgeometry of zirconia implants [47–65,67,70–75], while three studies evaluated the root-analog zirconia implants obtained by a three-dimensional scan [66,68,69]. However, Akça et al. and Pirker et al. reported the lowest marginal bone loss after two years (0.31 ± 0.24 and 0.5 ± 0.7 mm, respectively), in which Akça et al. used specially designed titanium–zirconia alloy implants [47], and Pirker et al. used specially designed root-analog zirconia implants with a micro-retention surface in a fresh extraction socket [69].

Table 2. Main characteristics of the clinical research included (Zir: Zirconia implant group, Tit: Titanium implant group, IF: impact factor, RCT: Randomized controlled trial).

Authors (Year) [Ref]	Journal			Cited By	Study Design	Patients (Implants)	Test	Control	Marginal Bone Loss (Mean ± SD)	Survival Rate	Failure	Follow Up
	Full Title	Rank	IF									
Pirker et al. (2008) [68]	International Journal of Oral and Maxillofacial Surgery	33	2.068	50	Case report	1 (1 Implant)	Microretention and sandblasted root-analogue zirconia implant	-	-	100%	-	2 years
Pirker et al. (2009) [69]	International Journal of Oral and Maxillofacial Surgery	33	2.068	58	Prospective Case Series	18 (18 Implants)	Microretention and sandblasted root-analogue zirconia implants	Sandblasted root-analogue zirconia implants	0.5 ± 0.7 mm	Test: 92% Control: 0%	Test: 1 implant Control: All implants (6)	2 years
Cannizzaro et al. (2010) [54]	European Journal of Implantology	-	-	69	Multicenter RCT	40 (40 Implants)	Immediate occlusal loading zirconia Implants	Immediate non-occlusal loading zirconia Implants	Test: 0.90 ± 0.48 mm Control: 0.72 ± 0.59 mm	88.50%	5 implants (12.5%): Test: 3 Implants Control: 2 Implants	1 year
Borgonovo et al. (2011) [51]	Minerva Stomatologica	-	-	21	Prospective Case Series	16 (26 Implants)	One-piece yttrium stabilized zirconia implants	-	-	96.16%	1 Implant osseointegration failure	2 years
Payer et al. (2012) [66]	Clinical Oral Implants Research	8	3.723	61	Prospective Case Series	20 (20 Implants)	One-piece zirconia implants	-	1.29 ± 0.73 mm	95%	1 Implant osseointegration failure	2 years
Akça et al. (2013) [47]	International Journal of Oral and Maxillofacial Implants	24	2.32	8	Prospective Case Series	23 (52 Implants)	-	-	0.32 ± 0.24 mm	100%	No failure	2 years
Borgonovo et al. (2013) [52]	Minerva Stomatologica	-	-	10	Prospective Case Series	6 (14 Implants)	One-piece yttrium stabilized zirconia implants	-	0.67 ± 0.51 mm	100%	No failure	4 years
Kohal et al. (2013) [59]	Journal of Clinical Periodontology	2	5.241	47	Prospective Case Series	28 (56 Implants)	One-piece yttria-stabilized tetragonal zirconia implants	-	1.95 ± 0.65	98.20%	1 Implant osseointegration failure	1 year
Osman et al. (2013) [63]	International Journal of Prosthodontics	61	1.49	6	Pilot study	4 (28 Implants)	One-piece zirconia implants for ball abutment	-	-	85.70%	4 Implants	1 year
Osman et al. (2014) [64]	Clinical Oral Implants Research	8	3.723	34	RCT	19 (129 Implants)	One-piece zirconia implants for ball-abutment	One-piece titanium implants for ball-abutment	Zir: 0.42 ± 0.40 Tit: 0.18 ± 0.47	Zir: 90.9%Tit: 95.8%	Zir: 21Implants (3 fractured) Tit: 10Implants	1 year
Becker et al. (2015) [50]	Clinical Oral Implants Research	8	3.723	15	Prospective Cohort Study	52 (52 Implants)	Two-piece zirconia implants	-	-	95.80%	2 Implants	2 years
Cionca et al. (2015) [55]	Clinical Oral Implants Research	8	3.723	43	Prospective Case Series	32 (49 Implants)	Two-piece zirconia implants	-	-	87%	6 Implants	1 year

Table 2. *Cont.*

Authors (Year) [Ref]	Journal			Cited By	Study Design	Patients (Implants)	Test	Control	Marginal Bone Loss (Mean ± SD)	Survival Rate	Failure	Follow Up
	Full Title	Rank	IF									
Jung et al. (2015) [56]	Clinical Oral Implants Research	8	3.723	27	Prospective Cohort Study	60 (71 Implants)	Immediate one-piece zirconia implants	-	0.78 ± 0.79 mm	98.30%	1 implant osseointegration failure	1 year
Payer et al. (2015) [67]	Clinical Oral Implants Research	8	3.723	41	RCT	22 (31 Implants)	Two-piece zirconia implants	Two-piece titanium implants	Zir: 1.48 ± 1.05 Tit: 1.43 ± 0.67	Zir: 93.3% Tit: 100%	Zir: 1 Implant Tit: No failure	2 years
Siddiqi et al. (2015) [70]	Clinical Implant Dentistry and Related Research	9	3.396	17	RCT	22 (150 Implants)	One-piece zirconia implants for ball-abutment	Titanium implants for one-piece ball-abutment	Zir: 2.23 ± 0.69 Tit: 1.59 ± 0.33	Zir: 67.6% Tit: 66.7%	Zir: 16 Implants Tit: 7 Implants	1 year
Spies et al. (2015) [73]	Journal of Dental Research	3	4.914	22	Prospective Cohort Study	40 (53 Implants)	One-piece alumina-toughened zirconia implant	-	0.79 ± 0.47 mm	94.2%	3 Implants osseointegration failure	3 years
Patankar et al. (2016) [65]	Journal of Maxillofacial and Oral Surgery	-	-	3	Case report	1 (1 Implant)	Microretention and sandblasted root-analogue zirconia implant	-	-	100%	-	1.5 year
Spies et al. (2016) [74]	Clinical Oral Implants Research	8	3.723	13	Prospective Cohort Study	27 (27 Implants)	Immediate one-piece alumina-toughened zirconia implant	-	0.77 ± 0.31 mm	88.90%	3 Implants osseointegration failure	1 year
Kniha et al. (2017) [58]	International Journal of Oral and Maxillofacial Surgery	33	2.068	9	Prospective Cohort Study	81 (105 Implants)	Zirconia implants	-	0.66 ± 0.33 mm	100%	No failure	3 years
Kniha et al. (2017) [57]	International Journal of Oral and Maxillofacial Surgery	33	2.068	9	Prospective Cohort Study	78 (82 Implants)	Immediate loading one-piece zirconia implants	Delayed one-piece zirconia implants	Immediate: 0.76 ± 1.13 mm Delayed: 0.83 ± 0.65 mm	Immediate: 100% Delayed: 100%	No failure	1 year
Spies et al. (2017) [71]	Journal of Dentistry	10	3.242	6	Prospective Case Series	60 (71 Implants)	One-piece zirconia oral implants	-	-	100%	No failure	3 years
Spies et al. (2017) [75]	Clinical Oral Implants Research	8	3.723	11	Prospective Case Series	13 (26 Implants)	One-piece zirconia implants	-	-	100%	No failure	5 years
Balmer et al. (2018) [49]	Clinical Oral Implants Research	8	3.723	11	Prospective Multicenter Cohort Study	60 (71 Implants)	One-piece immediate loading zirconia implants	-	0.70 ± 0.72 mm	98.50%	1 Implant osseointegration failure	3 years
Bormann et al. (2018) [53]	BMC Oral Health	38	1.911	7	Prospective Multicenter Cohort Study	44 (44 Implants)	Zirconia implants	-	0.97 ± 0.88 mm	97.50%	1 Implant	3 years
Kohal et al. (2018) [60]	Journal of Clinical Periodontology	2	5.241	5	Prospective Cohort Study	65 (65 Implants)	Immediate loading one-piece zirconia implants	-	1.45 ± 1.96 mm	90.80%	6 Implants	3 years

Table 2. *Cont.*

Authors (Year) [Ref]	Journal			Cited By	Study Design	Patients (Implants)	Test	Control	Marginal Bone Loss (Mean ± SD)	Survival Rate	Failure	Follow Up
	Full Title	Rank	IF									
Lorenz et al. (2019) [62]	Clinical Implant Dentistry and Related Research	9	3.396	4	Prospective Cohort Study	28 (83 Implants)	Zirconia implants	Natural teeth	1.2 ± 0.76 mm	100%	No failure one peri-implantitis resistant to therapies	7.8 years
Spies et al. (2019) [72]	Clinical Oral Implants Research	8	3.723	5	Prospective Multicenter Cohort Study	45 (45 Implants)	Zirconia implants	-	-	97.5 ± 2.47%.	Chipping (n = 19) occlusal roughness (n = 35)	5 years
Balmer et al. (2020) [48]	Clinical Oral Implants Research	8	3.723	4	Prospective Multicenter Cohort Study	60 (71 Implants)	Single crown one-piece zirconia implant	Multiple prostheses one-piece zirconia implant	0.7 ± 0.6 mm	98.4%	1 Implant	5 years
Koller et al. (2020) [61]	Clinical Oral Implants Research	8	3.723	0	Pilot RCT	22 (31 Implants)	Two-piece zirconia implants	Two-piece titanium implants	Zir: 1.38 ± 0.81 Tit: 1.17 ± 0.73 mm	Zir: 87.5%Tit: 93.3%	Zir: 2 Implants Tit: 1 Implant	6.67 years

3.3. Growth of Publications

In total, 29 clinical studies were published between 2008 and 2020, in which 19 papers (65.5%) were published in the last five years and ten papers published before 2015. The highest number of published studies was in 2015 (n = 6, 20.6%) followed by 2013 and 2017 (n = 4, 13.7% for each) (Figure 2).

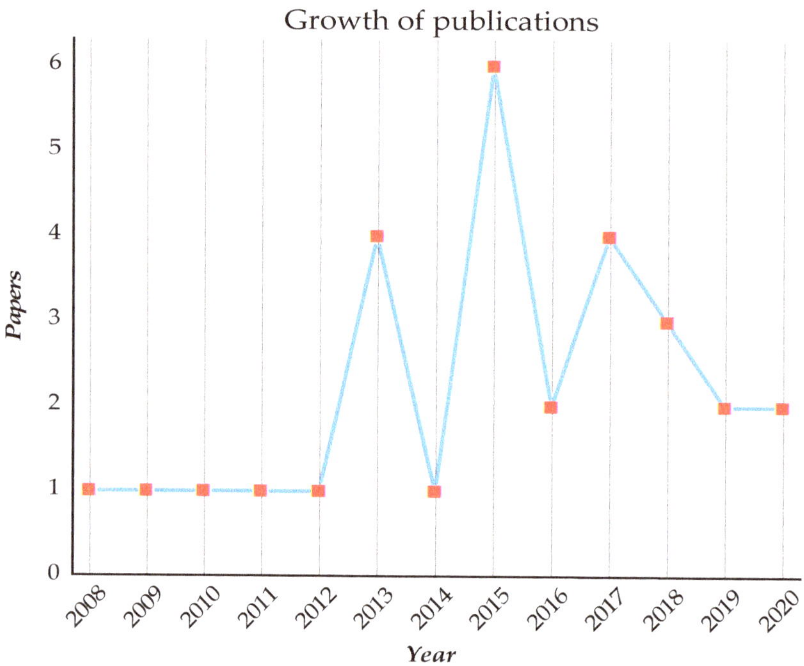

Figure 2. Publication trend of the clinical studies on the zirconia implants.

3.4. Journal of Publication

The clinical studies on the use of zirconia dental implants for oral rehabilitation were published across ten peer-reviewed journals. The journal with the largest number of publications was "Clinical Oral Implants Research" (n = 12, 41%), followed by "International Journal of Oral and Maxillofacial Surgery" (n = 4, 13.7%) (Figure 3).

The majority of publications were published in Q1 journals (n = 25, 86%), while the journal with the highest impact factor was "Journal of Clinical Periodontology" (IF = 5.241), which had two articles.

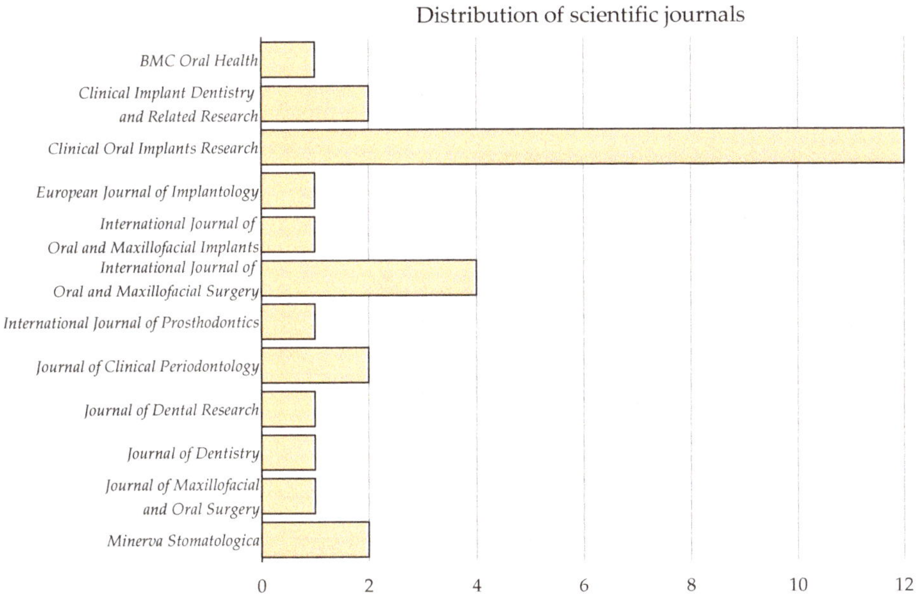

Figure 3. Contribution journals in clinical research on zirconia implants.

3.5. *Study Design and Level of Evidence*

All included studies were prospective, while the most common study design of clinical research on zirconia implants was cohort study (n = 12, 41.4%), followed by case series (n = 9, 31%), and RCT (n = 5, 17%). According to the hierarchy of evidence levels (Is) [77,78], the available evidence supporting the use of zirconia implants is 17% level II, 41.4% EL IV, and the remaining EL VI.

3.6. *Contribution of Countries and Institutions*

The majority of the studies originated from institutions in Germany (n = 14, 48.3%), followed by Switzerland, (n = 6, 20.7%), and Austria (n = 5, 17%), where the most productive institution was the Medical Center of University of Freiburg (n = 8, 27.6%), followed by the Center of Dental Medicine, University of Zürich (n = 5, 17%). While many of the included studies were funded, the most funding support for included research was provided by VITA Zahnfabrik—H. Rauter GmbH & Co. KG, Bad Säckingen, Germany (n = 5, 17%) (Table 3).

Table 3. Contribution of countries and institutions to clinical studies on zirconia implants.

Country	Institution	Study [Ref]	Funding
Germany	Universitätsklinikum Düsseldorf, Düsseldorf	Becker et al., 2015 [50]	ZV3 Zircon Vision GmbH, Wolfratshausen, Germany
	University Hospital Aachen, Aachen	Kniha et al., 2017 [57]	No Funding
	Friedrich-Alexander-University Erlangen-Nürnberg	Kniha et al., 2017 [58]	
	Johann-Wolfgang Goethe University, Frankfurt/Main	Lorenz et al., 2019 [62]	
	Hannover Medical School, Hannover	Bormann et al., 2018 [53]	Institut Straumann AG, Basel, Switzerland
	School of Dentistry, Albert-Ludwigs University, Freiburg	Kohal et al., 2013 [59]	Nobel Biocare AB, Göteborg, Sweden
	Medical Center of University of Freiburg, Freiburg	Kohal et al., 2018 [60]	
		Spies et al., 2015 [73]	Metoxit AG (Thayngen, Switzerland)
		Spies et al., 2016 [74]	
		Spies et al., 2017 [75]	Ivoclar Vivadent
Germany and Switzerland	Medical Center of University of Freiburg, Freiburg and Center of Dental Medicine, University of Zürich, Zürich	Spies et al., 2017 [71]	VITA Zahnfabrik—H. Rauter GmbH & Co. KG, Bad Säckingen, Germany
		Balmer et al., 2018 [49]	
		Spies et al., 2019 [72]	
		Balmer et al., 2020 [48]	
Switzerland	School of Dental Medicine, University of Geneva, Geneva	Cionca et al., 2015 [55]	Dentalpoint AG, Zürich, Switzerland
	Center of Dental Medicine, University of Zürich, Zürich	Jung et al., 2015 [56]	VITA Zahnfabrik—H. Rauter GmbH & Co. KG, Bad Säckingen, Germany
Austria	Alfred Kocher, Medical University Vienna, Vienna	Pirker et al., 2008 [68]	No Funding
		Pirker et al., 2009 [69]	
	School of Dentistry, Medical University Graz, Graz	Payer et al., 2012 [66]	Bredent medical GmbH, Senden, Germany
		Payer et al., 2015 [67]	Ziterion GmbH, Uffenheim, Germany
		Koller et al., 2020 [61]	
Italy	Private practice	Cannizzaro et al., 2010 [54]	Partially supported by Z-systems
	School of Dentistry, University of Milan, Milan	Borgonovo et al., 2011 [51]	Not reported
		Borgonovo et al., 2013 [52]	
New Zealand	Oral Implantology Research Group, Sir John Walsh Research Institute, School of Dentistry, University of Otago	Osman et al., 2013 [63]	Oral Implantology Research Group, Sir John Walsh Research Institute, School of Dentistry, University of Otago and Southern Implants
		Osman et al., 2014 [64]	
		Siddiqi et al., 2015 [70]	
India	BV Dental College and Hospital, Pune	Patankar et al., 2016 [65]	No Funding
Turkey	Faculty of Dentistry, Hacettepe University	Akça et al., 2013 [47]	No Funding

3.7. Bibliometric Assessment

A total of 29 articles with total citations$_{[Scopus]}$ ranged from 0 to 176 (mean 57.28 ± 42.18), while the number of citations$_{[Scopus]}$ received by each paper ranged from 0 to 69 (mean 21.3 ± 20). The top-cited study was the RCT of Cannizzaro et al. (2010) ($n_{[Scopus]}$ = 69) [54], followed by the prospective case series of Payer et al. (2012) ($n_{[Scopus]}$ = 61) [66], and Pirker et al. (2009) ($n_{[Scopus]}$ = 58) [69].

However, the author with the highest number of clinical research on zirconia implants was Kohal R.J. (n = 10), followed by Spies B.C. (n = 8) and Vach K. (n = 6), while the top-cited author of clinical studies on zirconia implants was Kohal R.J. ($n_{[WoS]}$ = 155), followed by Arnetzl G. ($n_{[WoS]}$ = 91), Koller M. ($n_{[WoS]}$ = 87), Payer M., and Jakse N. ($n_{[WoS]}$ = 86 for each) (Figures 4 and 5).

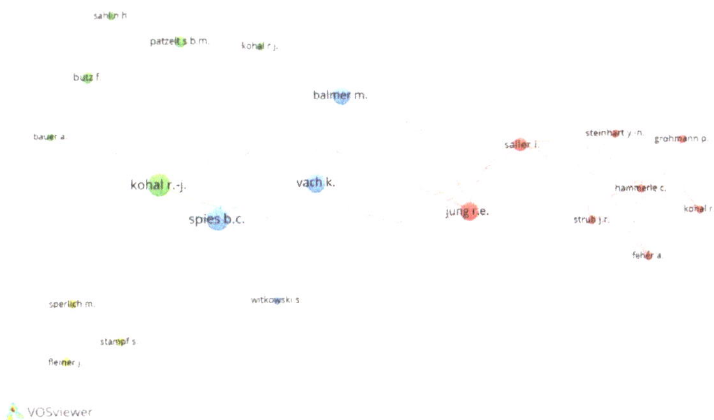

Figure 4. Network analysis of the authors with the largest number of clinical studies on zirconia implants.

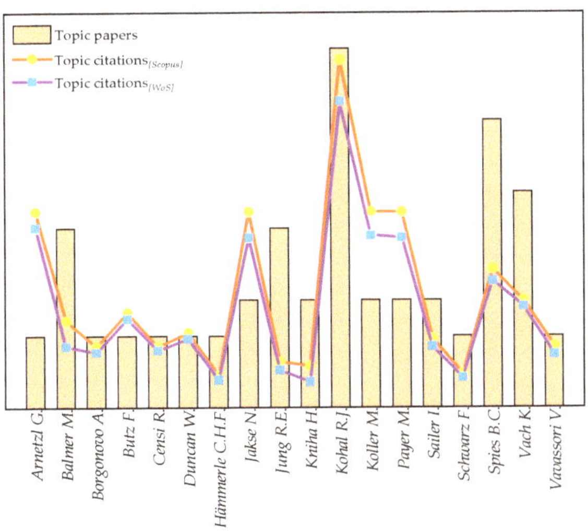

Figure 5. Bibliometric variables for authors with the highest number of topic papers.

The authors' H-index$_{Scopus}$ ranged from one to 79 (mean 22.67 ± 19.96), and the author with the most bibliometric characteristics was Hämmerle C.H.F., who had 364 publications (two of which were clinical studies on zirconia implants) with 8311 total citations and H-index$_{Scopus}$ = 79 (Table 4 and Figure 6).

Table 4. General bibliometric variables for authors with the largest number of topic papers.

Author	Topic Papers	Total Papers	Topic/Total Papers %	Topic Citations [Scopus]	Topic Citations [WoS]	Total Citations [Scopus]	Total Citation [WoS]	Topic/Total Citations % [Scopus-WoS]	H-index [Scopus]	H-index [WoS]
Kohal R.J.	10	109	9.17%	176	155	3053	2975	[5.76–5.21%]	30	29
Spies B.C.	8	47	17.02%	70	64	452	448	[15.49–14.29%]	14	14
Vach K.	6	73	8.22%	54	51	661	820	[8.17–6.21%]	15	16
Balmer M.	5	13	38.46%	44	31	256	130	[17.19–23.84%]	7	5
Jung R.E.	5	202	2.48%	23	19	8359	9126	[0.28–0.21%]	47	57
Jakse N.	3	70	4.29%	99	86	1114	981	[8.89–8.77%]	18	18
Kniha H.	3	28	10.71%	21	13	664	726	[3.16–1.779%]	11	13
Koller M.	3	9	33.33%	99	87	166	312	[59.64–27.88%]	5	5
Payer M.	3	48	6.25%	99	86	770	1484	[12.86–5.79%]	15	23
Sailer I.	3	113	2.65%	35	31	6027	5713	[0.58–0.54%]	34	33
Arnetzl G.	2	40	5.00%	99	91	437	183	[22.65–49.72%]	11	6
Butz F.	2	24	8.33%	48	45	1248	1441	[3.85–3.12%]	18	19
Censi R.	2	18	11.11%	31	29	169	95	[18.34–30.52%]	8	4
Duncan W.	2	87	2.30%	38	35	2163	1356	[1.76–2.58%]	20	19
Hämmerle C.H.F.	2	364	0.55%	16	14	18,311	16,032	[0.09–0.08%]	79	72
Schwarz F.	2	261	0.77%	17	15	9093	9169	[0.19–0.16%]	57	57
Vavassori V.	2	12	16.67%	31	27	136	96	[22.79–28.12%]	7	5
Borgonovo A.	2	51	3.92%	31	28	535	449	[5.79–6.23%]	12	11
Summary (Mean ± SD)	3.61 ± 2.33	87.17 ± 96.37	10.07 ± 0.11%	57.28 ± 42.18	50.39 ± 37.68	2978.56 ± 4750.05	2863.11 ± 4386.90	-	22.67 ± 19.96	22.56 ± 20.14

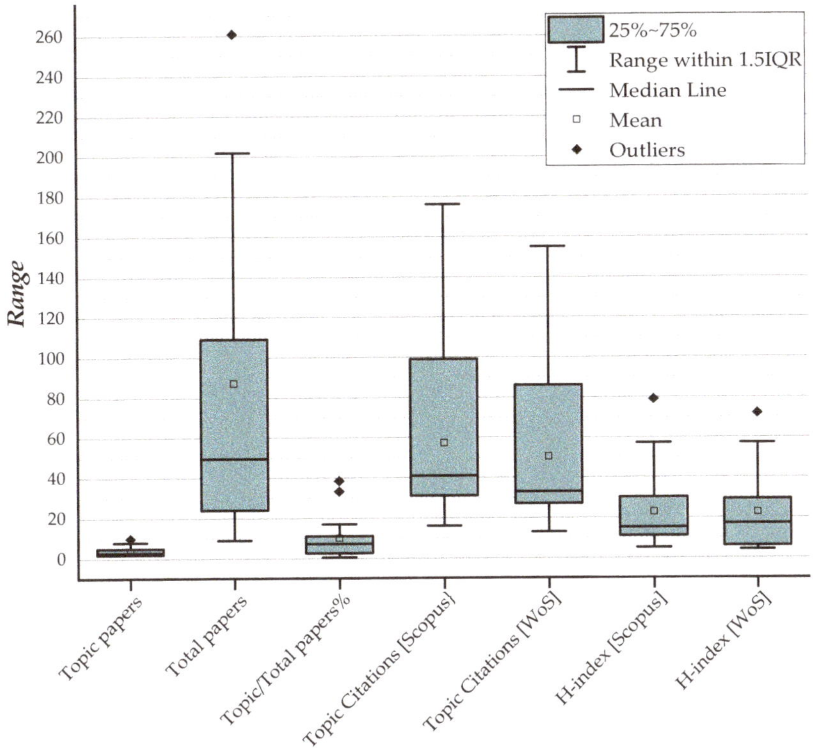

Figure 6. Box plots summarize the bibliometric variables of the authors with the largest number of studies.

4. Discussion

The present study carried out a bibliometric evaluation of clinical research on zirconia implant rehabilitation, highlighting the significant heterogeneity of the included studies, which revealed considerable variations in methodology, technical approaches, follow-up, and control group involvement. Our findings indicate that there is a trend for zirconia implants in oral rehabilitation as there has been an increase in about 180% of the studies published in the last five years.

The included studies reported a survival rate for zirconia implants ranging from 87% to 100% with follow-up periods from one to 7.8 years, while the least survival rate reported in RCT by Siddiqi et al. was 67.6% after one-year follow-up (i.e., 16 zirconia implants failed out of 68) [70]. This RCT aimed to study the effectiveness of zirconia vs. titanium implants restored with one-piece ball-abutment in mandibular and maxillary overdentures, while this high decrease in the survival rate was for both groups (i.e., 67.6% for zirconia implants and 66.7% for titanium implants); the outcomes of maxillary rehabilitation were worse than the mandible, while no mechanical fractures of the fixtures were reported [70].

Although one-piece and two-piece zirconia implants have been evaluated, the lower marginal bone loss and higher survival rates were observed in studies of one-piece zirconia implant rehabilitation on a single tooth or three element prosthetic rehabilitation [59,61]. However, the studies did not report any differences in the marginal bone loss and survival rate between the single crown and the fixed multiple zirconia implant recovery, while the prosthetic connection appears to have no apparent effect on these parameters [48]. Additionally, Lorenz et al. showed no significant difference in marginal bone loss with a total of 83 zirconia implants compared to natural teeth after 7.8 years of function [47], and the marginal bone loss was similar in the other studies, which was less than 1 mm in the first year and stabilized in subsequent functional loading [47–49,52,56–58,64,69,73,74]. Moreover, the prospective study by Kniha et al. contained the largest sample size of the included studies involving 81 patients with 105 implants for fixed rehabilitation, who reported a significant decrease of 0.66 ± 0.30 mm with a survival rate of 100% after three years [58].

However, the most common complication (70%) was the failure of implant osseointegration as 17 studies reported a loss of at least one implant in the first six months [48–51,53–56,59–61,63,64,66,67, 69,70,72–74].

As previously reported for titanium dental implant threads, microgeometry appears to have a significant effect on the osseointegration of zirconia implants [79,80], whereas a more retentive surface resulted in an increased survival rate compared to a sandblasted surface only [68,69].

Although all clinical research included in this analysis was screened and selected from the Scopus and Web of Science databases, which may avoid restriction in each database [39,81], our investigation has further limitations. First, the year of publication, which is a reliable indicator of the number of citations received, as older papers receive more citations than recent publications because there is more time to cite them, regardless of their impact [82,83]. Second, open access policies have a significant influence on the citations received in the evaluated papers [84–86], as a result, we found large heterogeneity in Topic/Total Citations% and co-authors' H-index.

5. Conclusions

This was the first study highlighting bibliometric output of clinical research and co-authors in the field of zirconia dental implants and shows a strong interest in the development of research into the clinical application of zirconia dental implants, as evidenced by the increase in the number of scientific papers published in the last ten years.

Author Contributions: Conceptualization, F.L. and A.S.; Methodology, F.L. and A.G.A.K.; Software, F.L.; Validation, F.L. and A.S.; Formal analysis, F.L. and A.G.A.K.; Investigation, A.S., F.L. and B.R.; Data curation, F.L., A.G.A.K. and A.S.; Writing—original draft preparation, F.L. and A.S.; Writing—review and editing, F.L., A.G.A.K. and S.N.; Visualization, A.S., B.R. and I.F.; Supervision, A.S. and I.F. All authors have read and agreed to the published version of the manuscript.

Funding: This research received no external funding.

Conflicts of Interest: Authors declare no conflicts of interest.

References

1. Albrektsson, T.; Wennerberg, A. On osseointegration in relation to implant surfaces. *Clin. Implant Dent. Relat. Res.* **2019**, *21* (Suppl. 1), 4–7. [CrossRef]
2. Buser, D.; Janner, S.F.M.; Wittneben, J.-G.; Brägger, U.; Ramseier, C.A.; Salvi, G.E. 10-Year Survival and Success Rates of 511 Titanium Implants with a Sandblasted and Acid-Etched Surface: A Retrospective Study in 303 Partially Edentulous Patients. *Clin. Implant Dent. Relat. Res.* **2012**, *14*, 839–851. [CrossRef] [PubMed]
3. Degidi, M.; Piattelli, A. A 7-year Follow-up of 93 Immediately Loaded Titanium Dental Implants. *J. Oral Implantol.* **2005**, *31*, 25–31. [CrossRef] [PubMed]
4. Scarano, A.; Inchingolo, F.; Murmura, G.; Traini, T.; Piattelli, A.; Lorusso, F. Three-Dimensional Architecture and Mechanical Properties of Bovine Bone Mixed with Autologous Platelet Liquid, Blood, or Physiological Water: An In Vitro Study. *Int. J. Mol. Sci.* **2018**, *19*. [CrossRef] [PubMed]
5. Barros, R.R.M.; Degidi, M.; Novaes, A.B.; Piattelli, A.; Shibli, J.A.; Iezzi, G. Osteocyte Density in the Peri-Implant Bone of Immediately Loaded and Submerged Dental Implants. *J. Periodontol.* **2009**, *80*, 499–504. [CrossRef]
6. Gehrke, S.; Mazón, P.; Del Fabbro, M.; Tumedei, M.; Aramburú Júnior, J.; Pérez-Díaz, L.; De Aza, P. Histological and Histomorphometric Analyses of Two Bovine Bone Blocks Implanted in Rabbit Calvaria. *Symmetry* **2019**, *11*, 641. [CrossRef]

7. Scarano, A.; Carinci, F.; Lorusso, F.; Festa, F.; Bevilacqua, L.; Santos de Oliveira, P.; Maglione, M. Ultrasonic vs Drill Implant Site Preparation: Post-Operative Pain Measurement Through VAS, Swelling and Crestal Bone Remodeling: A Randomized Clinical Study. *Mater. Basel* **2018**, *11*, 2516. [CrossRef]
8. Scarano, A.; Crincoli, V.; Di Benedetto, A.; Cozzolino, V.; Lorusso, F.; Podaliri Vulpiani, M.; Grano, M.; Kalemaj, Z.; Mori, G.; Grassi, F.R. Bone Regeneration Induced by Bone Porcine Block with Bone Marrow Stromal Stem Cells in a Minipig Model of Mandibular "Critical Size" Defect. *Stem. Cells Int.* **2017**, *2017*, 9082869. [CrossRef]
9. Scarano, A.; De Oliveira, P.S.; Traini, T.; Lorusso, F. Sinus Membrane Elevation with Heterologous Cortical Lamina: A Randomized Study of a New Surgical Technique for Maxillary Sinus Floor Augmentation without Bone Graft. *Mater. Basel* **2018**, *11*, 1457. [CrossRef]
10. Piattelli, A.; Scarano, A.; Piattelli, M. Detection of alkaline and acid phosphatases around titanium implants: A light microscopical and histochemical study in rabbits. *Biomaterials.* **1995**, *16*, 1333–1338. [CrossRef]
11. Comuzzi, L.; Tumedei, M.; Piattelli, A.; Iezzi, G. Short vs. Standard Length Cone Morse Connection Implants: An In Vitro Pilot Study in Low Density Polyurethane Foam. *Symmetry* **2019**, *11*, 1349. [CrossRef]
12. Tumedei, M.; Savadori, P.; Del Fabbro, M. Synthetic Blocks for Bone Regeneration: A Systematic Review and Meta-Analysis. *Int. J. Mol. Sci.* **2019**, *20*, 4221. [CrossRef] [PubMed]
13. Afrashtehfar, K.I.; Del Fabbro, M. Clinical performance of zirconia implants: A meta-review. *J. Prosthet. Dent.* **2020**, *123*, 419–426. [CrossRef] [PubMed]
14. Hanawa, T. Zirconia versus titanium in dentistry: A review. *Dent. Mater. J.* **2020**, *39*, 24–36. [CrossRef] [PubMed]
15. Hashim, D.; Cionca, N.; Courvoisier, D.S.; Mombelli, A. A systematic review of the clinical survival of zirconia implants. *Clin. Oral Investig.* **2016**, *20*, 1403–1417. [CrossRef]
16. Cantore, S.; Mirgaldi, R.; Ballini, A.; Coscia, M.F.; Scacco, S.; Papa, F.; Inchingolo, F.; Dipalma, G.; De Vito, D. Cytokine gene polymorphisms associate with microbiogical agents in periodontal disease: Our experience. *Int. J. Med. Sci.* **2014**, *11*, 674–679. [CrossRef]
17. Noronha Oliveira, M.; Schunemann, W.V.H.; Mathew, M.T.; Henriques, B.; Magini, R.S.; Teughels, W.; Souza, J.C.M. Can degradation products released from dental implants affect peri-implant tissues? *J. Periodontal Res.* **2017**, *53*, 1–11. [CrossRef]
18. Noumbissi, S.; Scarano, A.; Gupta, S. A Literature Review Study on Atomic Ions Dissolution of Titanium and Its Alloys in Implant Dentistry. *Mater. Basel* **2019**, *12*, 368. [CrossRef]
19. Ottria, L.; Lauritano, D.; Andreasi Bassi, M.; Palmieri, A.; Candotto, V.; Tagliabue, A.; Tettamanti, L. Mechanical, chemical and biological aspects of titanium and titanium alloys in implant dentistry. *J. Biol. Regul. Homeost. Agents* **2018**, *32* (Suppl. 1), 81–90.
20. Rodrigues, D.C.; Valderrama, P.; Wilson, T.G.; Palmer, K.; Thomas, A.; Sridhar, S.; Adapalli, A.; Burbano, M.; Wadhwani, C. Titanium Corrosion Mechanisms in the Oral Environment: A Retrieval Study. *Mater. Basel* **2013**, *6*, 5258–5274. [CrossRef]
21. Scarano, A.; Piattelli, M.; Caputi, S.; Favero, G.A.; Piattelli, A. Bacterial Adhesion on Commercially Pure Titanium and Zirconium Oxide Disks: An In Vivo Human Study. *J. Periodontol.* **2004**, *75*, 292–296. [CrossRef] [PubMed]
22. Mihatovic, I.; Golubovic, V.; Becker, J.; Schwarz, F. Bone tissue response to experimental zirconia implants. *Clin. Oral Investig.* **2016**, *21*, 523–532. [CrossRef] [PubMed]
23. Scarano, A.; Di Carlo, F.; Quaranta, M.; Piattelli, A. Bone Response to Zirconia Ceramic Implants: An Experimental Study in Rabbits. *J. Oral Implantol.* **2003**, *29*, 8–12. [CrossRef]
24. Agustín-Panadero, R.; Serra-Pastor, B.; Roig-Vanaclocha, A.; Fons-Font, A.; Solá-Ruiz, M.F. Fracture resistance and the mode of failure produced in metal-free crowns cemented onto zirconia abutments in dental implants. *PLoS ONE* **2019**, *14*, e0220551. [CrossRef] [PubMed]
25. Bethke, A.; Pieralli, S.; Kohal, R.-J.; Burkhardt, F.; von Stein-Lausnitz, M.; Vach, K.; Spies, B.C. Fracture Resistance of Zirconia Oral Implants In Vitro: A Systematic Review and Meta-Analysis. *Mater. Basel* **2020**, *13*, 562. [CrossRef] [PubMed]
26. Sailer, I.; Asgeirsson, A.G.; Thoma, D.S.; Fehmer, V.; Aspelund, T.; Özcan, M.; Pjetursson, B.E. Fracture strength of zirconia implant abutments on narrow diameter implants with internal and external implant abutment connections: A study on the titanium resin base concept. *Clin. Oral Implant. Res.* **2018**, *29*, 411–423. [CrossRef]

27. Cao, Y.; Yu, C.; Wu, Y.; Li, L.; Li, C. Long-Term Survival and Peri-Implant Health of Titanium Implants with Zirconia Abutments: A Systematic Review and Meta-Analysis. *J. Prosthodont.* **2019**, *28*, 883–892. [CrossRef]
28. Fanali, S.; Tumedei, M.; Pignatelli, P.; Inchingolo, F.; Pennacchietti, P.; Pace, G.; Piattelli, A. Implant primary stability with an osteocondensation drilling protocol in different density polyurethane blocks. *Comput. Methods Biomech. Biomed. Eng.* **2020**, 1–7. [CrossRef]
29. Fujiwara, S.; Kato, S.; Bengazi, F.; Urbizo Velez, J.; Tumedei, M.; Kotsu, M.; Botticelli, D. Healing at implants installed in osteotomies prepared either with a piezoelectric device or drills: An experimental study in dogs. *Oral Maxillofac. Surg.* **2020**. [CrossRef]
30. Kermanshah, H.; Geramy, A.; Ebrahimi, S.F.; Bitaraf, T. IPS-Empress II inlay-retained fixed partial denture reinforced with zirconia bar: Three-dimensional finite element andin-vitrostudies. *Acta Odontol. Scand.* **2012**, *70*, 569–576. [CrossRef]
31. Kotsu, M.; Urbizo Velez, J.; Bengazi, F.; Tumedei, M.; Fujiwara, S.; Kato, S.; Botticelli, D. Healing at implants installed from ~ 70- to <10-Ncm insertion torques: An experimental study in dogs. *Oral Maxillofac. Surg.* 2020. [CrossRef]
32. Scarano, A.; Di Carlo, F.; Piattelli, A. Effect of sterilization and cleansing on implantology drills: Zirconia vs steel. *Ital. Oral Surg.* **2008**, *3*, 61–72.
33. Scarano, A.; Valbonetti, L.; Marchetti, M.; Lorusso, F.; Ceccarelli, M. Soft Tissue Augmentation of the Face with Autologous Platelet-Derived Growth Factors and Tricalcium Phosphate. Microtomography Evaluation of Mice. *J Craniofac. Surg.* **2016**, *27*, 1212–1214. [CrossRef] [PubMed]
34. Scarano, A.; Piattelli, A.; Quaranta, A.; Lorusso, F. Bone Response to Two Dental Implants with Different Sandblasted/Acid-Etched Implant Surfaces: A Histological and Histomorphometrical Study in Rabbits. *BioMed Res. Int.* **2017**, *2017*, 8724951. [CrossRef] [PubMed]
35. Ibrahim, G.M.; Carter Snead, O.; Rutka, J.T.; Lozano, A.M. The most cited works in epilepsy: Trends in the "Citation Classics". *Epilepsia* **2012**, *53*, 765–770. [CrossRef]
36. Parker, J.N.; Lortie, C.; Allesina, S. Characterizing a scientific elite: The social characteristics of the most highly cited scientists in environmental science and ecology. *Scientometrics* **2010**, *85*, 129–143. [CrossRef]
37. Park, K.M.; Park, B.S.; Park, S.; Yoon, D.Y.; Bae, J.S. Top-100 cited articles on headache disorders: A bibliometric analysis. *Clin. Neurol. Neurosurg.* **2017**, *157*, 40–45. [CrossRef]
38. Lorusso, F.; Inchingolo, F.; Scarano, A. Scientific Production in Dentistry: The National Panorama through a Bibliometric Study of Italian Academies. *BioMed Res. Int.* **2020**, *2020*, 3468303. [CrossRef]
39. Jayaratne, Y.S.N.; Zwahlen, R.A. The evolution of dental journals from 2003 to 2012: A bibliometric analysis. *PLoS ONE* **2015**, *10*, e0119503. [CrossRef]
40. Pommer, B.; Valkova, V.; Ubaidha Maheen, C.; Fürhauser, L.; Rausch-Fan, X.; Seeman, R. Scientific Interests of 21st Century Clinical Oral Implant Research: Topical Trend Analysis. *Clin. Implant Dent. Relat. Res.* **2015**, *18*, 850–856. [CrossRef]
41. Gutiérrez-Vela, M.M.; Díaz-Haro, A.; Berbel-Salvador, S.; Lucero-Sánchez, A.; Robinson-García, N.; Cutando-Soriano, A. Bibliometric analysis of research on regenerative periodontal surgery during the last 30 years. *J. Clin. Exp. Dent.* **2012**, *4*, e112–e118. [CrossRef] [PubMed]
42. Tetè, S.; Zizzari, V.L.; De Carlo, A.; Lorusso, F.; Di Nicola, M.; Piattelli, A.; Gherlone, E.; Polimeni, A. Characterizing scientific production of Italian Oral Surgery professionals through evaluation of bibliometric indices. *Ann. Di Stomatol.* **2014**, *5*, 23. [CrossRef]
43. O'Brien, B.C.; Harris, I.B.; Beckman, T.J.; Reed, D.A.; Cook, D.A. Standards for Reporting Qualitative Research. *Acad. Med.* **2014**, *89*, 1245–1251. [CrossRef] [PubMed]
44. Liberati, A.; Altman, D.G.; Tetzlaff, J.; Mulrow, C.; Gøtzsche, P.C.; Ioannidis, J.P.A.; Clarke, M.; Devereaux, P.J.; Kleijnen, J.; Moher, D. The PRISMA statement for reporting systematic reviews and meta-analyses of studies that evaluate health care interventions: Explanation and elaboration. *PLoS Med.* **2009**, *6*, e1000100. [CrossRef]
45. Van Eck, N.J.; Waltman, L. Software survey: VOSviewer, a computer program for bibliometric mapping. *Scientometrics* **2010**, *84*, 523–538. [CrossRef]
46. Van Eck, N.J.; Waltman, L. VOSviewer manual. *Leiden Univeristeit Leiden* **2013**, *1*, 1–53.
47. Akça, K.; Cavusoglu, Y.; Uysal, S.; Cehreli, M.C. A Prospective, Open-Ended, Single-Cohort Clinical Trial on Early Loaded Titanium-Zirconia Alloy Implants in Partially Edentulous Patients: Up-to-24-Month Results. *Int. J. Oral Maxillofac. Implant.* **2013**, *28*, 573–578. [CrossRef]

48. Balmer, M.; Spies, B.C.; Kohal, R.J.; Hämmerle, C.H.F.; Vach, K.; Jung, R.E. Zirconia implants restored with single crowns or fixed dental prostheses: 5-year results of a prospective cohort investigation. *Clin. Oral Implant. Res.* **2020**, *31*, 452–462. [CrossRef]
49. Balmer, M.; Spies, B.C.; Vach, K.; Kohal, R.-J.; Hämmerle, C.H.F.; Jung, R.E. Three-year analysis of zirconia implants used for single-tooth replacement and three-unit fixed dental prostheses: A prospective multicenter study. *Clin. Oral Implant. Res.* **2018**, *29*, 290–299. [CrossRef]
50. Becker, J.; John, G.; Becker, K.; Mainusch, S.; Diedrichs, G.; Schwarz, F. Clinical performance of two-piece zirconia implants in the posterior mandible and maxilla: A prospective cohort study over 2 years. *Clin. Oral Implant. Res.* **2015**, *28*, 29–35. [CrossRef]
51. Borgonovo, A.; Censi, R.; Dolci, M.; Vavassori, V.; Bianchi, A.; Maiorana, C. Use of endosseous one-piece yttrium-stabilized zirconia dental implants in premolar region: A two-year clinical preliminary report. *Minerva. Stomatol.* **2011**, *60*, 229–241. [PubMed]
52. Borgonovo, A.; Vavassori, V.; Censi, R.; Calvo, J.; Re, D. Behavior of endosseous one-piece yttrium stabilized zirconia dental implants placed in posterior areas. *Minerva. Stomatol.* **2013**, *62*, 247–257. [PubMed]
53. Bormann, K.-H.; Gellrich, N.-C.; Kniha, H.; Schild, S.; Weingart, D.; Gahlert, M. A prospective clinical study to evaluate the performance of zirconium dioxide dental implants in single-tooth edentulous area: 3-year follow-up. *BMC Oral Health* **2018**, *18*, 181. [CrossRef] [PubMed]
54. Cannizzaro, G.; Torchio, C.; Felice, P.; Leone, M.; Esposito, M. Immediate occlusal versus non-occlusal loading of single zirconia implants. A multicentre pragmatic randomised clinical trial. *Eur. J. Oral Implant.* **2010**, *3*, 111–120.
55. Cionca, N.; Müller, N.; Mombelli, A. Two-piece zirconia implants supporting all-ceramic crowns: A prospective clinical study. *Clin. Oral Implant. Res.* **2015**, *26*, 413–418. [CrossRef]
56. Jung, R.E.; Grohmann, P.; Sailer, I.; Steinhart, Y.-N.; Fehér, A.; Hämmerle, C.; Strub, J.R.; Kohal, R. Evaluation of a one-piece ceramic implant used for single-tooth replacement and three-unit fixed partial dentures: A prospective cohort clinical trial. *Clin. Oral Implant. Res.* **2015**, *27*, 751–761. [CrossRef]
57. Kniha, K.; Kniha, H.; Möhlhenrich, S.C.; Milz, S.; Hölzle, F.; Modabber, A. Papilla and alveolar crest levels in immediate versus delayed single-tooth zirconia implants. *Int. J. Oral Maxillofac. Surg.* **2017**, *46*, 1039–1044. [CrossRef]
58. Kniha, K.; Schlegel, K.A.; Kniha, H.; Modabber, A.; Hölzle, F.; Kniha, K. Evaluation of peri-implant bone levels and soft tissue dimensions around zirconia implants—A three-year follow-up study. *Int. J. Oral Maxillofac. Surg.* **2017**, *47*, 492–498. [CrossRef]
59. Kohal, R.-J.; Patzelt, S.B.M.; Butz, F.; Sahlin, H. One-piece zirconia oral implants: One-year results from a prospective case series. 2. Three-unit fixed dental prosthesis (FDP) reconstruction. *J. Clin. Periodontol.* **2013**, *40*, 553–562. [CrossRef] [PubMed]
60. Kohal, R.-J.; Spies, B.C.; Bauer, A.; Butz, F. One-piece zirconia oral implants for single-tooth replacement: Three-year results from a long-term prospective cohort study. *J. Clin. Periodontol.* **2018**, *45*, 114–124. [CrossRef] [PubMed]
61. Koller, M.; Steyer, E.; Theisen, K.; Stagnell, S.; Jakse, N.; Payer, M. Two-piece zirconia versus titanium implants after 80 months: Clinical outcomes from a prospective randomized pilot trial. *Clin. Oral Implant. Res.* **2020**, *31*, 388–396. [CrossRef] [PubMed]
62. Lorenz, J.; Giulini, N.; Hölscher, W.; Schwiertz, A.; Schwarz, F.; Sader, R. Prospective controlled clinical study investigating long-term clinical parameters, patient satisfaction, and microbial contamination of zirconia implants. *Clin. Implant Dent. Relat. Res.* **2019**, *21*, 263–271. [CrossRef] [PubMed]
63. Osman, R.B.; Payne, A.; Duncan, W.; Ma, S. Zirconia implants supporting overdentures: A pilot study with novel prosthodontic designs. *Int. J. Prosthodont.* **2013**, *26*, 277–281. [CrossRef]
64. Osman, R.B.; Swain, M.V.; Atieh, M.; Ma, S.; Duncan, W. Ceramic implants (Y-TZP): Are they a viable alternative to titanium implants for the support of overdentures? A randomized clinical trial. *Clin. Oral Implant. Res.* **2014**, *25*, 1366–1377. [CrossRef] [PubMed]
65. Patankar, A.; Kshirsagar, R.; Patankar, S.; Pawar, S. Immediate, non submerged root analog zirconia implant in single rooted tooth replacement: Case report with 2 years follow up. *J. Maxillofac. Oral Surg.* **2016**, *15*, 270–273. [CrossRef]

66. Payer, M.; Arnetzl, V.; Kirmeier, R.; Koller, M.; Arnetzl, G.; Jakse, N. Immediate provisional restoration of single-piece zirconia implants: A prospective case series-results after 24 months of clinical function. *Clin. Oral Implant. Res.* **2012**, *24*, 569–575. [CrossRef]
67. Payer, M.; Heschl, A.; Koller, M.; Arnetzl, G.; Lorenzoni, M.; Jakse, N. All-ceramic restoration of zirconia two-piece implants—A randomized controlled clinical trial. *Clin. Oral Implant. Res.* **2015**, *26*, 371–376. [CrossRef]
68. Pirker, W.; Kocher, A. Immediate, non-submerged, root-analogue zirconia implant in single tooth replacement. *Int. J. Oral Maxillofac. Surg.* **2008**, *37*, 293–295. [CrossRef]
69. Pirker, W.; Kocher, A. Immediate, non-submerged, root-analogue zirconia implants placed into single-rooted extraction sockets: 2-year follow-up of a clinical study. *Int. J. Oral Maxillofac. Surg.* **2009**, *38*, 1127–1132. [CrossRef]
70. Siddiqi, A.; Kieser, J.A.; De Silva, R.K.; Thomson, W.M.; Duncan, W.J. Soft and Hard Tissue Response to Zirconia versus Titanium One-Piece Implants Placed in Alveolar and Palatal Sites: A Randomized Control Trial. *Clin. Implant Dent. Relat. Res.* **2015**, *17*, 483–496. [CrossRef]
71. Spies, B.C.; Balmer, M.; Jung, R.E.; Sailer, I.; Vach, K.; Kohal, R.-J. All-ceramic, bi-layered crowns supported by zirconia implants: Three-year results of a prospective multicenter study. *J. Dent.* **2017**, *67*, 58–65. [CrossRef] [PubMed]
72. Spies, B.C.; Balmer, M.; Jung, R.E.; Sailer, I.; Vach, K.; Kohal, R.J. All-ceramic single crowns supported by zirconia implants: 5-year results of a prospective multicenter study. *Clin. Oral Implant. Res.* **2019**, *30*, 466–475. [CrossRef] [PubMed]
73. Spies, B.C.; Balmer, M.; Patzelt, S.B.M.; Vach, K.; Kohal, R.J. Clinical and Patient-reported Outcomes of a Zirconia Oral Implant: Three year Results of a Prospective Cohort Investigation. *J. Dent. Res.* **2015**, *94*, 1385–1391. [CrossRef] [PubMed]
74. Spies, B.C.; Sperlich, M.; Fleiner, J.; Stampf, S.; Kohal, R.-J. Alumina reinforced zirconia implants: 1-year results from a prospective cohort investigation. *Clin. Oral Implant. Res.* **2016**, *27*, 481–490. [CrossRef] [PubMed]
75. Spies, B.C.; Witkowski, S.; Vach, K.; Kohal, R.J. Clinical and patient-reported outcomes of zirconia-based implant fixed dental prostheses: Results of a prospective case series 5 years after implant placement. *Clin. Oral Implant. Res.* **2017**, *29*, 91–99. [CrossRef] [PubMed]
76. Borgonovo, A.-E.; Fabbri, A.; Vavassori, V.; Censi, R.; Maiorana, C. Multiple teeth replacement with endosseous one-piece yttrium-stabilized zirconia dental implants. *Med. Oral Patol. Oral Cir. Bucal.* **2012**, *17*, e981–e987. [CrossRef] [PubMed]
77. Greenhalgh, T. How to read a paper: Getting your bearings (deciding what the paper is about). *BMJ* **1997**, *315*, 243–246. [CrossRef]
78. Guyatt, G.H.; Sackett, D.L.; Sinclair, J.C.; Hayward, R.; Cook, D.J.; Cook, R.J.; Bass, E.; Gerstein, H.; Haynes, B.; Holbrook, A. Users' guides to the medical literature: IX. A method for grading health care recommendations. *JAMA* **1995**, *274*, 1800–1804. [CrossRef]
79. Scarano, A.; Crocetta, E.; Quaranta, A.; Lorusso, F. Influence of the thermal treatment to address a better osseointegration of Ti6Al4V dental implants: Histological and histomorphometrical study in a rabbit model. *BioMed Res. Int.* **2018**. [CrossRef]
80. Scarano, A.; Degidi, M.; Perrotti, V.; Degidi, D.; Piattelli, A.; Iezzi, G. Experimental Evaluation in Rabbits of the Effects of Thread Concavities in Bone Formation with Different Titanium Implant Surfaces. *Clin. Implant Dent. Relat. Res.* **2013**, *16*, 572–581. [CrossRef]
81. Van Eck, N.J.; Waltman, L. Accuracy of citation data in Web of Science and Scopus. *arXiv* **2019**, arXiv:1906.07011.
82. Feijoo, J.F.; Limeres, J.; Fernández-Varela, M.; Ramos, I.; Diz, P. The 100 most cited articles in dentistry. *Clin. Oral Investig.* **2014**, *18*, 699–706. [CrossRef] [PubMed]
83. Ugolini, D.; Neri, M.; Cesario, A.; Bonassi, S.; Milazzo, D.; Bennati, L.; Lapenna, L.M.; Pasqualetti, P. Scientific production in cancer rehabilitation grows higher: A bibliometric analysis. *Supportive Care Cancer* **2012**, *20*, 1629–1638. [CrossRef] [PubMed]
84. Davis, P.M. Open access, readership, citations: A randomized controlled trial of scientific journal publishing. *FASEB J.* **2011**, *25*, 2129–2134. [CrossRef] [PubMed]

85. Gargouri, Y.; Hajjem, C.; Larivière, V.; Gingras, Y.; Carr, L.; Brody, T.; Harnad, S. Self-selected or mandated, open access increases citation impact for higher quality research. *PLoS ONE* **2010**, *5*, e13636. [CrossRef] [PubMed]
86. Hua, F.; Shen, C.; Walsh, T.; Glenny, A.-M.; Worthington, H. Open Access: Concepts, findings, and recommendations for stakeholders in dentistry. *J. Dent.* **2017**, *64*, 13–22. [CrossRef] [PubMed]